Design of
Reconfigurable Antennas
Using Graph Models

Synthesis Lectures on Antennas

Editor

Constantine A. Balanis, *Arizona State University*

Synthesis Lectures on Antennas will publish 50- to 100-page publications on topics that include both classic and advanced antenna configurations. Each lecture covers, for that topic, the fundamental principles in a unified manner, develops underlying concepts needed for sequential material, and progresses to the more advanced designs. State-of-the-art advances made in antennas are also included. Computer software, when appropriate and available, is included for computation, visualization and design. The authors selected to write the lectures are leading experts on the subject who have extensive background in the theory, design and measurement of antenna characteristics.

The series is designed to meet the demands of 21st century technology and its advancements on antenna analysis, design and measurements for engineers, scientists, technologists and engineering managers in the fields of wireless communication, radiation, propagation, communication, navigation, radar, RF systems, remote sensing, and radio astronomy who require a better understanding of the underlying concepts, designs, advancements and applications of antennas.

Design of Reconfigurable Antennas Using Graph Models
Joseph Costantine, Youssef Tawk, and Christos G. Christodoulou
2013

Meta-Smith Charts and Their Potential Applications
Danai Torrungrueng
2010

Generalized Transmission Line Method to Study the Far-zone Radiation of Antennas under a Multilayer Structure
Xuan Hui Wu, Ahmed A. Kishk, and Allen W. Glisson
2008

Narrowband Direction of Arrival Estimation for Antenna Arrays
Jeffrey Foutz, Andreas Spanias, and Mahesh K. Banavar
2008

Multiantenna Systems for MIMO Communications
Franco De Flaviis, Lluis Jofre, Jordi Romeu, and Alfred Grau
2008

Design of Reconfigurable Antennas Using Graph Models

Joseph Costantine, Youssef Tawk, and Christos G. Christodoulou

ISBN: 978-3-031-00412-4   paperback
ISBN: 978-3-031-01540-3   ebook

DOI 10.1007/978-3-031-01540-3

A Publication in the Springer series
SYNTHESIS LECTURES ON ANTENNAS

Lecture #11
Series Editor: Constantine A. Balanis, *Arizona State University*
Series ISSN
Synthesis Lectures on Antennas
Print 1932-6076   Electronic 1932-6084

Design of Reconfigurable Antennas Using Graph Models

Joseph Costantine
American University of Beirut

Youssef Tawk
The University of New Mexico

Christos G. Christodoulou
The University of New Mexico

SYNTHESIS LECTURES ON ANTENNAS #11

ABSTRACT

This lecture discusses the use of graph models to represent reconfigurable antennas. The rise of antennas that adapt to their environment and change their operation based on the user's request hasn't been met with clear design guidelines. There is a need to propose some rules for the optimization of any reconfigurable antenna design and performance. Since reconfigurable antennas are seen as a collection of self-organizing parts, graph models can be introduced to relate each possible topology to a corresponding electromagnetic performance in terms of achieving a characteristic frequency of operation, impedance, and polarization. These models help designers understand reconfigurable antenna structures and enhance their functionality since they transform antennas from bulky devices into mathematical and software accessible models. The use of graphs facilitates the software control and cognition ability of reconfigurable antennas while optimizing their performance.

This lecture also discusses the reduction of redundancy, complexity and reliability of reconfigurable antennas and reconfigurable antenna arrays. The full analysis of these parameters allows a better reconfigurable antenna implementation in wireless and space communications platforms. The use of graph models to reduce the complexity while preserving the reliability of reconfigurable antennas allow a better incorporation in applications such as cognitive radio, MIMO, satellite communications, and personal communication systems. A swifter response time is achieved with less cost and losses. This lecture is written for individuals who wish to venture into the field of reconfigurable antennas, with a little prior experience in this area, and learn how graph rules and theory, mainly used in the field of computer science, networking, and control systems can be applied to electromagnetic structures. This lecture will walk the reader through a design and analysis process of reconfigurable antennas using graph models with a practical and theoretical outlook.

KEYWORDS

reconfigurable antennas, reconfigurable systems, graph theory, graph models, switches, redundancy, complexity, switch reliability, reconfigurable antenna arrays, switch failure

To our families

Contents

CHAPTER 1

Introduction to Reconfigurable Antennas

1.1 INTRODUCTION

The concept of reconfigurable antennas can be dated back to a 1983 patent of D. Schaubert [1]. In 1999, the Defense Advanced Research Projects Agency (DARPA), in the United States, sponsored an initiative under the name "Reconfigurable Aperture Program (RECAP)," in order to investigate reconfigurable antennas and their potential applications [2]. Since then, reconfigurable antennas have been used in broadband communication, cognitive radio, MIMO systems, and other applications.

Reconfiguring an antenna can be achieved by changing its frequency, polarization, or radiation characteristics. Most techniques used to achieve reconfigurability in antennas redistribute the antenna currents and thus alter the electromagnetic fields of the antenna's effective aperture.

Reconfigurable patch antennas are the most widely developed reconfigurable type antennas due to their ease of fabrication and integration into small electronic devices such as cell phones and laptops. Usually, a typical reconfigurable patch antenna consists of a number of separate metalized regions which lie on a plane and are connected together through switches or tuning elements. By dynamically controlling the state of the switches, different metalized sections can be brought in contact together, thereby altering the radiation performance of the total antenna.

When designing reconfigurable antennas, RF engineers must address three challenging questions [3, 4]:

1. Which reconfigurable property (e.g., frequency, radiation pattern, or polarization) needs to be modified?

2. How is the reconfigurable antenna's topology reconfigured to address question 1?

3. Which reconfiguration technique addresses questions 1 and 2 and minimizes the negative effects on the antenna radiation/impedance characteristics?

1.2 RECONFIGURATION TECHNIQUES AND CLASSIFICATIONS OF RECONFIGURABLE ANTENNAS

The major types of reconfiguration techniques that can be used to implement reconfigurable antennas are indicated in Fig. 1.1 [3]. Antennas based on RF-MEMS [5–10], PIN diodes [11–22], and varactors [23–29] that redirect their surface currents are called "electrically reconfigurable." Antennas that make use of photoconductive switching elements are called "optically reconfigurable" antennas [30–34]. A description of the operation of the switches is summarized in Table 1.1 [35–38]. Reconfigurable antennas can also be achieved by mechanically altering the structure of the antenna [39–42]. These are called "physically reconfigurable" antennas. Finally, reconfigurable antennas can also be implemented using smart materials such as ferrites and liquid crystals [43–46].

In general, reconfigurable antennas can be categorized into four groups [47] that are divided based on their reconfigurable operation as shown in Fig. 1.2. Group 1 includes antennas that exhibit reconfigurability in their operation or notch frequency. Group 2 includes antennas that exhibit reconfigurable radiation patterns. Group 3 includes antennas with reconfigurable polarization, while group 4 includes antennas with multiple properties such as frequency reconfiguration with polarization diversity.

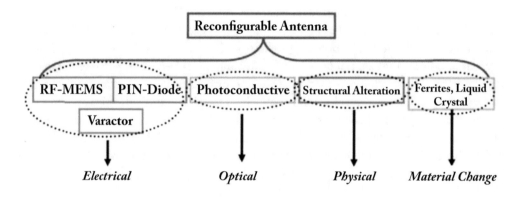

Figure 1.1: Various techniques adopted to achieve reconfigurable antennas [3].

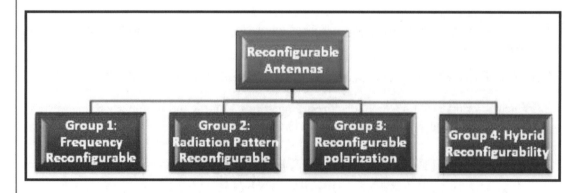

Figure 1.2: Categorization of Reconfigurable Antennas.

Table 1.1: Different type of switches used in the design of electrically and optically reconfigurable antennas [3]

<u>RF MEMS:</u> They use mechanical movement to achieve a short circuit or an open circuit in a surface current path of an antenna structure. The forces required for the mechanical movement can be obtained using electrostatic, magnetostatic, piezoelectric, or thermal designs.
<u>PIN Diodes:</u> They operate in two modes. The "ON" state, where the diode is forward biased and the "OFF" state, where the diode is not biased or reverse biased.
<u>Varactors:</u> They consist of a p-n junction diode. As the bias voltage applied to the diode is varied, the varactor capacitance is going to be changed. Typical values are from tens to hundreds of picofarads.
<u>Photoconductive Elements:</u> The movement of electrons from the valence band to the conduction band allows the switch to go from "OFF" state to "ON" state. This is achieved by illuminating the switch by light of appropriate wavelength from a laser diode.

The corresponding reconfigurability for each of the four groups can be obtained by a change in the antenna current distribution, a change in the feeding network, a change in the antenna physical structure, or a change in the antenna radiating edges. In fact, the change in one parameter in the antenna characteristics can affect the other parameters. Therefore, an antenna engineer needs to be careful during the design process to analyze all the antenna characteristics simultaneously in order to achieve the required reconfigurability.

While reconfigurable antennas represent potential candidates for future RF front-end in wireless communication applications, there is a cost for adding tuning ability to the antenna behavior [3]. This cost can be linked to different parameters as summarized below [3]:

1. Design of the required biasing networks to activate/deactivate the switching elements which add complexity to the antenna structure

2. Increase in power consumption due to the addition of active components

3. Generation of harmonics and inter modulation products

4. Need for fast tuning in the antenna radiation characteristics to insure an accurate functioning of the system.

In the next section we limit our discussion to "electrically reconfigurable antennas" since in this book our main focus is on using graphs to model electrically reconfigurable antennas.

1.3   ELECTRICALLY RECONFIGURABLE ANTENNAS

An electrically reconfigurable antenna relies on electronic switching components (RF-MEMs, PIN diodes, or varactors) to redistribute its surface currents, and alter the antenna radiating structure topology and/or radiating edges. The integration of switches into the antenna structure makes it easier for designers to achieve the desired reconfigurable functionality.

The versatility of this type of reconfigurable antennas has attracted many researchers despite the numerous issues surrounding the use of such switching techniques. These issues include the non-linearity effects of switches, and the negative effect of the associated bias lines used to control the state of the switching components on the antenna radiation pattern.

Three different examples of electrically reconfigurable antennas are described. Each example discusses the use of a different reconfiguration technique to reach the corresponding function.

1.3.1   RECONFIGURABLE ANTENNAS BASED ON RF-MEMS

The antenna shown in Fig. 1.3 is a reconfigurable rectangular type of spiral antenna with a set of embedded RF-MEMs switches. The antenna structure consists of five sections that are connected with four RF-MEMS switches. The spiral arm is increased following the right-hand direction to yield right-hand circular polarization for the radiated field. Based on the status of the integrated RF-MEMS, the antenna can change its radiation beam direction [5].

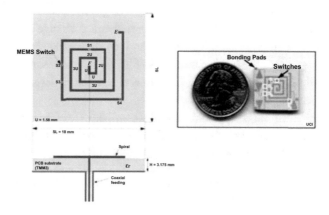

Figure 1.3: (Left) a radiation pattern reconfigurable antenna; (right) fabricated prototype with the biasing line [5].

1.3.2   RECONFIGURABLE ANTENNAS BASED ON PIN DIODES

A frequency and pattern reconfigurable antenna based on PIN diodes is discussed in [11]. The activation of the switches is automated via a field programmable gate array (FPGA). The metal patch composed of a main midsection and four surrounding smaller sections are shown in Fig. 1.4(a). The variations in configuration are achieved through individually controllable switches, each implemented as a PIN diode. The antenna tunes its operating frequency/radiation pattern according to the four switch combinations. Fig. 1.4(b) shows how an FPGA can be used to control the activation and deactivation of these switches.

1.3.3   RECONFIGURABLE ANTENNAS BASED ON VARACTORS

A wideband dual sided Vivaldi antenna is tuned by incorporating a varactor-based reconfigurable band pass filter into its 50 ohms microstrip feeding line [48]. The band-pass filter's microstrip line is composed of three sections: Two outer sections and a hexagonal slot that is etched in the center of the middle microstrip line section. A varactor is incorporated inside the hexagonal slot to achieve a variable capacitive connection between the two terminals in the slot of the middle section. The whole antenna system is now reconfigurable through the varactor tuned filter that is a part of the antenna's microstrip feeding line [48]. The whole system (antenna + filter) is called "Filtenna." The filtenna top and bottom layers are shown in Fig. 1.5. The antenna has a partial ground which is the ground plane of the filter of dimensions 30mm x 30mm. The top layer constitutes the first side of the antenna radiating surface as well as the feeding line where the reconfigurable filter is located. On the bottom layer of the design resides the ground plane of the filter connected to the second radiating part of the Vivaldi antenna. The change in the biasing voltage tunes the varactor's capacitance which in turns tunes the Filtenna's operating frequency [48].

1.4   RECONFIGURABLE ANTENNAS APPLICATIONS AND REQUIREMENTS

Reconfigurable antennas are designed to be implemented on various platforms which cover various wireless services that are spanned over a wide frequency range. In particular, reconfigurable antennas are proposed for higher efficiencies in various implementations that involve cognition and continuous adaptation to the environment such as in cognitive radio and MIMO systems.

A cognitive radio system requires a frequency reconfigurable antenna that can communicate across a channel by changing its frequency of operation based on the constant monitoring of the channel spectrum. This system is able to continuously monitor gaps (white spaces) in the finite frequency spectrum occupied by other wireless systems, and then dynamically change its transmit/receive characteristics to operate within these unused frequency bands. The monitoring of the wireless spectrum is the key in cognitive radio since the spectrum can be idle for 90% of the time [49]. Therefore, we should differentiate in such systems between a primary user that owns the spectrum and a secondary user that wants to access the spectrum whenever it is idle [50].

This capability requires an "Ultra Wide Band (UWB) sensing antenna" that continuously monitors the wireless channel searching for unused carrier frequencies, and a "reconfigurable transmit/receive antenna" to perform the data transfer [51–55] as shown in Figure 1.6.

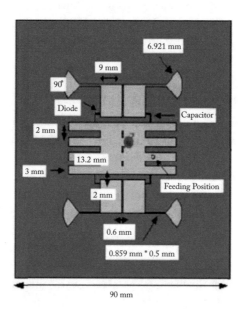

(a)

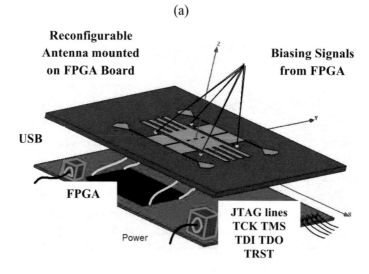

Figure 1.4: (a) Frequency reconfigurable antenna [11]; (b) FPGA controlling reconfigurable antenna [47].

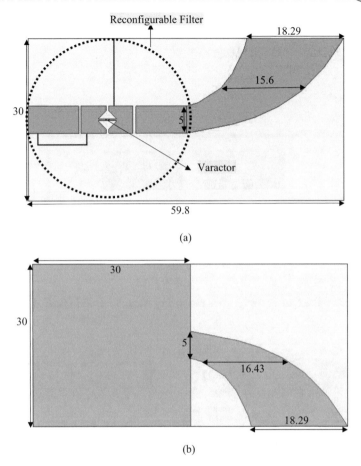

Figure 1.5: (a) The filtenna top layer; (b) bottom layer [48].

A Multiple Input Multiple Output (MIMO) system employs multiple antennas at both the transmitter and the receiver front-ends. The advantage from using such configurations is that different information can be sent simultaneously, thereby increasing the communication spectral efficiency in a multipath environment. According to the varying channel conditions and user's need, a MIMO system can adjust the modulation level, coding rate, and the transmission signaling schemes. Radiation pattern/polarization reconfigurable antennas add an additional degree of freedom in a MIMO environment and thus improve the system performance. The use of this type of antennas increases significantly the capacity by allowing the selection between different pattern diversity and polarization configurations. Reconfigurable antenna arrays are also an attractive solution for MIMO systems to maintain good communication links, especially for handheld devices where space is an important constraint [56–59]. An example is shown in Figure 1.7 where two reconfigurable antennas, activated via PIN diodes, can be used to form a MIMO system that operates over a wide bandwidth with notch frequency reconfigurability [60].

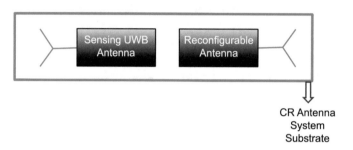

Figure 1.6: Basic block diagram of a cognitive radio that includes a wideband antenna and a reconfigurable antenna [3].

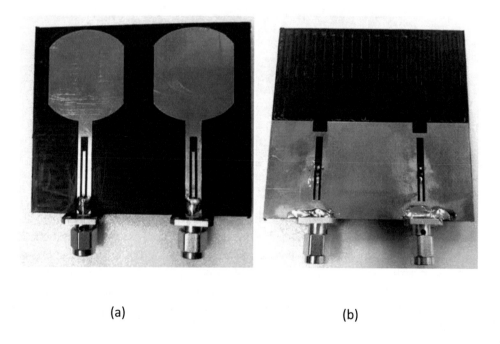

Figure 1.7: MIMO systems based on PIN diode switch activated reconfigurable antennas [60]. (a) Top layer, (b) Bottom layer.

Reconfigurable antennas can also be used in modern space applications. In such cases, it is required to be able to reconfigure the antenna radiation pattern to serve a new coverage zone, limit fading in rainy areas, and maintain high data rate at as many frequency bands as possible [61–65].

Most of the advanced wireless communications' applications (cognitive radio, MIMO, Space communications) require highly efficient software controlled dynamic antennas. These antennas that can be reconfigured using software and based on users' requests have to be highly reliable as well as able to achieve the required functions. Therefore, an antenna designer proposing a new reconfigurable antenna design for an advanced application has to present a design that has minimal losses in its structure and operation. The antenna has to exhibit software accessibility as well as be able to be controlled using programmable controllers. Most importantly, a designer must insure that the proposed antenna is able to continuously operate under unforeseen circumstances; thus a reliability study is required from reconfigurable antenna designers.

In the next chapters, we propose using graph models as a tool to transform antennas from bulky structures into software accessible devices that can be optimized and software controlled. A design methodology using graph modeling techniques is proposed to remove redundant components from reconfigurable antenna structures without affecting the systems' reliability while reducing their complexity and computational requirements. Reconfigurable antenna arrays' properties are also discussed, and a new technique for the detection and correction of switch failures in large array structures is also presented at the end of this book.

CHAPTER 2

Graph Modeling Reconfigurable Antennas

2.1 INTRODUCTION

Reconfigurable antennas have been designed by integrating various reconfiguration components (switches, actuators, etc.), into an antenna structure. The reconfiguration process has been executed by activating and deactivating the reconfiguring elements to connect or disconnect parts of an antenna as well as to alter the antenna's radiated fields and current distributions. These mechanisms help achieve user controlled tuning in the antenna performance. In fact, the rise of reconfigurable antennas is based on the fact that these antennas are software controlled and can be automated easily. Thus, a benefit exists in improving the software control process of such antennas. Such improvement can be achieved by increasing the antennas' reconfiguration speed as well as employing various algorithms into the antennas' automation process to optimize their purpose and achieve self-operation. The main objective is to design antennas that adapt to their environment and tune their operation as quickly and as efficiently as possible. Antennas have to be transformed from their "bulky" state as front end devices into software-accessible system components that can be automated and optimized accordingly.

The software control of antennas requires their modeling in a mathematical form. Several model possibilities have been investigated, out of which "Graph Models" have been widely and heavily used in areas such as computer science, control systems, and networking [66]. By definition, graphs are mathematical illustrations of various systems; they are symbolic representations of relationships between different points in a system. They are mathematical tools used to model real life situations, in order to organize them and improve their status [67]. A graph can also be a description of a communication protocol; in particular, a suitably designed graph can precisely describe and direct the changing network topology of a self-organizing system [67].

Graphs are also investigated in [68–71] to model the topologies of self-assembled systems such as self-assembly robots and organize their communication protocols. Different parts [68–71] self-organize based on different rules and various conditions. For example, reconfigurable antennas can be considered as "self-assembly" systems.

Since reconfigurable antennas can be seen as a collection of self-organizing parts, graphs can be used to model this type of antennas. Graph rules are introduced to relate each possible topology to a corresponding electromagnetic performance in terms of achieving a characteristic frequency of operation, radiation characteristics, or polarization. These rules help designers understand reconfigurable antenna structures and their operation. Various algorithms can be incorporated such as search algorithms and shortest path algorithms which facilitate the design, automation, and optimization process of reconfigurable antennas. In this chapter, graph models and their terminologies are introduced as well as guidelines for the modeling of reconfigurable antennas are set.

2.2   INTRODUCTION TO GRAPHS

In this section, we introduce graph models and define some of their properties that play an integral part in the modeling process of reconfigurable antennas.

2.2.1   WHAT IS A GRAPH?

A graph is a drawing representation of a complex system in a certain situation. A graph models a certain system into various points or nodes called vertices that may be connected together by lines called edges [72–77]. Each vertex represents a certain component or node of the system, and an edge represents a certain connection that may exist between any two components. An edge may have identical vertices at its end points, i.e., it is possible to have a vertex joined to itself by an edge; such an edge is called a loop [72–77]. Let's consider the following example:

Three sportsmen are playing a game of "catch the ball." The players pass the ball to each other at different turns without any preference or direction. We assume that the ball is always caught by the various players and never dropped down. In order to construct this scenario by a graph model, the three players are represented first by three vertices called {A, B, C}. These vertices are connected by edges representing the action of throwing the ball by one player and catching the ball by another. There is an edge present between A and B, B and C, as well as between A and C. The graph model of this scenario is shown in Fig. 2.1. This graph can be changed as soon as any of the players drop the ball which interrupts the game.

2.2.2   THE PROPERTIES OF A GRAPH

A graph can be either directed or undirected. A directed graph is a diagram where the edges have a certain non-reciprocal direction. The directions of the different edges indicate the flow of operation of a certain system. For example, a system composed of non-reciprocal components can be modeled using a directed graph. A graph where the edges don't have a determined and specific direction is called undirected. This graph indicates that the flow of operation of a system can occur in any direction and is not specified.

For example, we assume that the game of "catch the ball" discussed in Section 2.2.1 has a different set of rules. In this version player B can't throw the ball, but only catches it from A and C, player A doesn't catch the ball but only throws it to both B and C, and player C can catch the ball from A and throw it to B.

In this case the graph representing this game is definitely directed. Directed edges are drawn from A to B, A to C, and from C to B as shown in Fig. 2.2.

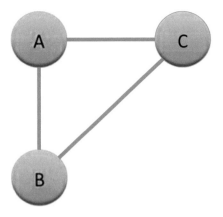

Figure 2.1: A graph representation of "Catch the Ball" game between three players.

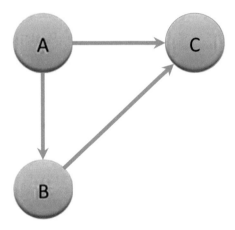

Figure 2.2: A directed graph.

Vertices represent physical entities. The edges between any two vertices in the graph represent the presence of a function resulting from connecting the physical entities. Edges may have weights associated with them to represent costs or benefits that are to be minimized or maximized. For example, if a capacitor is connecting two end points of a system and these end points are represented by two vertices in a graph, then the edge connecting these two vertices has a weight equal to the capacitance of that capacitor. Fig. 2.3 shows an example of an undirected as well as a weighted directed graph.

An empty graph is a graph with no edges. An edge of a graph is incident with its end vertex. Two edges that are incident with the same vertex are said to be adjacent. The degree of a vertex is the number of edges incident with that vertex, counting each loop twice. A vertex is called odd or even depending on whether its degree is odd or even [72].

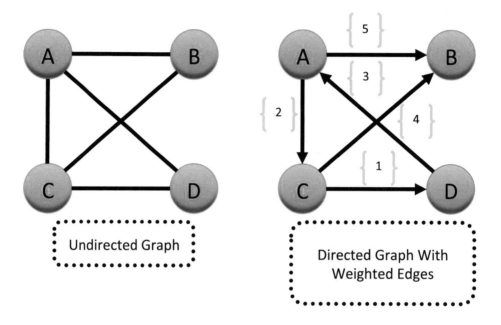

Figure 2.3: An example of an undirected as well as directed graph with weighted edges [47].

2.2.3   THE ADJACENCY MATRIX REPRESENTATION OF A GRAPH

The adjacency-matrix representation of a graph G, consisting of N vertices numbered $1, 2, \ldots, |N|$, is the $|M| \times |N|$ matrix A where each element a_{ij} of the matrix satisfies [66]:

$$a_{ij} = \begin{cases} 1 & \text{if there is an edge between vertex } i \text{ and vertex } j \\ 0 & \text{if there is no edge between vertex } i \text{ and vertex } j \end{cases} \tag{2.1}$$

The adjacency matrix of the graph shown in Fig. 2.1 is represented by the matrix A_1 below. The three vertices shown in the graph of Fig. 2.1 (A, B, C) correspond to the vertices (1, 2, 3) respectively. The edges between them are represented in the matrix A_1 following Eq. (2.1)

$$A_1 = \begin{bmatrix} 0 & 1 & 1 \\ 1 & 0 & 1 \\ 1 & 1 & 0 \end{bmatrix}$$

The adjacency-matrix representation can also be used for weighted graphs. The corresponding weights in a graph appear in the adjacency matrix. For example, the weight of a certain edge between two vertices 1 and 2 is simply stored as the entry in row 1 and column 2 of the adjacency matrix. If an edge does not exist, we will use the value "0" in the adjacency matrix to indicate the lack of an edge. For example the adjacency matrix of the graph shown in Fig. 2.3 can be written as below:

$$A_{undirected} = \begin{bmatrix} 0 & 1 & 1 & 1 \\ 1 & 0 & 1 & 0 \\ 1 & 1 & 0 & 1 \\ 1 & 0 & 1 & 0 \end{bmatrix} \qquad A_{directed\ with\ weights} = \begin{bmatrix} 0 & 5 & 2 & 0 \\ 0 & 0 & 0 & 0 \\ 0 & 4 & 0 & 1 \\ 3 & 0 & 0 & 0 \end{bmatrix}$$

2.2.4 WALKS AND PATHS IN A GRAPH

A walk w in a graph G from vertex V_0 to V_K is defined as the finite sequence:

$$w = v_0 e_1 v_1 e_2 v_2 \dots v_{k-1} e_k v_k \tag{2.2}$$

whose terms are alternatively vertices and edges such that, for $1 \le i \le k$, the edges e_i has ends v_{i-1} and v_i. Thus, each edge e_i is immediately preceded and succeeded by the two vertices with which they are incident. The vertex v_0 in the above walk is called the origin of walk W, while v_k is called the terminus of W [72]. For example, in Fig. 2.3 the walk from vertex A to vertex D in the directed graph is:

$$W_{A \to D} = A e_{A \to C} C e_{C \to D} D \tag{2.3}$$

If the vertices $v_0, v_1, \dots, v_k$ of the walk W are distinct then W is called a Path. For example, $W_{A \to D}$ in Eq. (2.3) is considered a path [72]. The weight of a path is defined as the sum of the weights of its constituent edges. The weight of the path $W_{A \to D}$ can be calculated as:

$$w\,(W_{A \to D}) = 2 + 1 = 3$$

In some cases it is useful to find the shortest path connecting two vertices. This notion is used in graph algorithms in order to optimize a certain function. The shortest path distance in a non-weighted graph is defined as the minimum number of edges in any path from vertex s to vertex **v**; otherwise if the graph is weighted then the shortest path corresponds to the least sum of weights in a particular path.

2.3 RULES AND GUIDELINES FOR GRAPH MODELING RECONFIGURABLE ANTENNAS

There are several ways to graph model reconfigurable antennas. Some rules for the graph modeling of reconfigurable antennas are set herein. These rules are not unique but they are required for our analysis process [47]. We set constraints for each rule in order to facilitate the graph modeling process. These constraints explain how to graph model each specific case of reconfigurable antennas. Herein, an antenna is called a multi-part antenna if it is composed of an array of identical or different elements (triangular, rectangular parts). Otherwise, it is called a single-part antenna. For example, a microstrip antenna whose patch contains three separate parts connected by switches is called a multi-part antenna; while a microstrip antenna whose patch is composed only of one radiating part is called a single-part antenna [47].

Rule 1: A multi-part antenna connected with switches is modeled as a weighted undirected graph. This graph consists of a vertex for each antenna part and connects those vertices with undirected weighted edges, wherever the different parts of the antenna have a physical connection through switches.

Constraints:

The connection between two parts has a distinctive angular direction. The designer defines a reference axis that represents the direction that the majority of parts have in relation to each other or with a main part.

The connections between the parts are represented by the edges. The edges' weights represent the angles that the connections make in relation to the reference axis. A weight $W = 1$ is assigned to an edge representing a connection that has an angle $0°$ or $180°$ in relation to the reference axis. Otherwise, a weight $W = 2$ is assigned to the edge forming any other angle with the reference axis as shown in Eq. (2.4).

$$W_{ij} = P_{ij} \qquad\qquad (2.4)$$
$$\text{where} \quad P_{ij} = \begin{cases} 1 & A_{ij} = 0° \text{ or } 180° \\ 2 & \text{otherwise} \end{cases}$$

where A_{ij} represents the angle that the connection i, j form with the reference axis.

Example 2.1

Let's take the antenna shown in Fig. 2.4 [78] and model it by a graph following rule 1. The basic antenna is a $130°$ balanced bowtie. Each bowtie arm consists of six triangular patches that are connected together via switches. Therefore, this antenna can be considered as a multi-part antenna.

Following rule 1, the vertices in the graph model represent the triangular patches. The graph model of this antenna consists of 12 vertices $(T_1, T_1', \ldots T_6, T_6')$. The edges connecting these vertices represent the connection of the corresponding triangles by MEMS switches. The reference axis for this antenna is taken along the left side of the upper part and along the right side of the lower part as shown in Fig. 2.4. Based on this reference axis direction and position, the edges appearing between $(T_1$ and $T_2)$, $(T_2$ and $T_4)$, $(T_1'$ and $T_3')$, and $(T_3'$ and $T_6')$ are collinear with the reference axis. The corresponding weights should be $W = 1$ according to Eq. (2.4) while the weights for all other edges should be set to $W = 2$. The adjacency matrix A containing all the weights is shown below:

$$A = \begin{bmatrix} 0 & 1 & 2 & 0 & 0 & 0 & 1 & 0 & 0 & 0 & 0 & 0 \\ 1 & 0 & 2 & 1 & 2 & 0 & 0 & 0 & 0 & 0 & 0 & 0 \\ 2 & 2 & 0 & 0 & 2 & 2 & 0 & 0 & 0 & 0 & 0 & 0 \\ 0 & 1 & 0 & 0 & 2 & 0 & 0 & 0 & 0 & 0 & 0 & 0 \\ 0 & 2 & 2 & 2 & 0 & 2 & 0 & 0 & 0 & 0 & 0 & 0 \\ 0 & 0 & 2 & 0 & 2 & 0 & 0 & 0 & 0 & 0 & 0 & 0 \\ 1 & 0 & 0 & 0 & 0 & 0 & 0 & 2 & 1 & 0 & 0 & 0 \\ 0 & 0 & 0 & 0 & 0 & 0 & 2 & 0 & 2 & 2 & 2 & 0 \\ 0 & 0 & 0 & 0 & 0 & 0 & 1 & 2 & 0 & 0 & 2 & 1 \\ 0 & 0 & 0 & 0 & 0 & 0 & 0 & 2 & 0 & 0 & 2 & 0 \\ 0 & 0 & 0 & 0 & 0 & 0 & 0 & 2 & 2 & 2 & 0 & 2 \\ 0 & 0 & 0 & 0 & 0 & 0 & 0 & 0 & 1 & 0 & 2 & 0 \end{bmatrix}$$

It is essential to note that since T_1 and T_1' are always connected, their graph representations should always consist of a non-weighted edge for all the different switch combinations. Fig. 2.5 presents the graph model of the antenna for two cases. Case 1: all switches are OFF, and Case 2: All switches along the reference axis are ON.

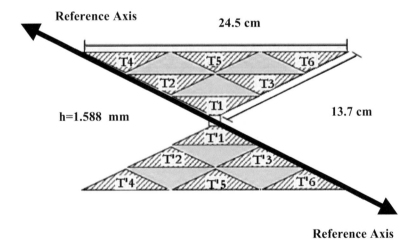

Figure 2.4: The antenna structure in [78].

Example 2.2

In this example we consider the antenna shown in Fig. 2.6 [79]. Reconfiguration is achieved by turning ON or OFF various antenna sections using switches, to change the active length of the assembled monopole structure. Therefore, the graph modeling of this antenna follows rule 1

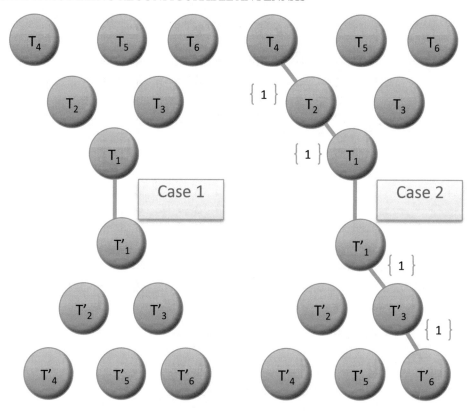

Figure 2.5: Graph model of two different antenna configurations in [78] using rule 1 for the case where all switches are OFF (Case 1) and the case where only the switches along the reference axis are ON (Case 2).

where the different parts are the vertices and the edges represent the connection of these parts by the activation of the different switches that connect the different parts. The reference axis is taken along the antenna radiating structure. Thus, the weight of all edges should be set to 1. The antenna structure can achieve three different configurations: P_0 and T_1 connected via switch S_1, P_0, P_1, and P_2 are connected via switches S_1 and S_2 as well as P_0, P_1, P_2, and P_3 are, respectively, connected via switches S_1, S_2, and S_3. Three different graph models can be built for the antenna as shown in Fig. 2.7. The adjacency matrix containing all the edges' weights is represented below:

$$A = \begin{bmatrix} 0 & 1 & 0 & 0 \\ 1 & 0 & 1 & 0 \\ 0 & 1 & 0 & 1 \\ 0 & 0 & 1 & 0 \end{bmatrix}.$$

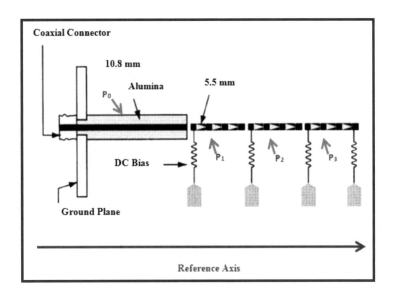

Figure 2.6: The antenna structure in [79].

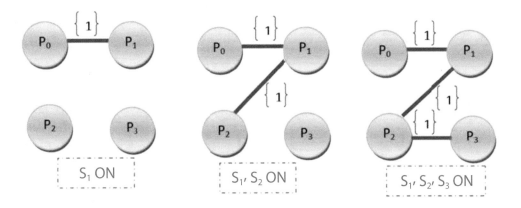

Figure 2.7: Antenna graph model.

Rule 2: A single part antenna with switches bridging over slots is modeled as a non-weighted undirected graph. This graph consists of a vertex for every switch end-point, and an undirected edge connects those vertices wherever switches are activated.

Constraints:

In the case of switches bridging multiple slots in one antenna structure, the graph model takes into consideration one slot at a time.

Example 2.3

As an example, let's take the antenna shown in Fig. 2.8 [80]. This antenna is a triangular patch antenna with two slots. The authors suggested five switches to bridge each slot in order to achieve the required functions.

The graph modeling of this antenna following rule 2 is shown in Fig. 2.9, for all possible connections. The vertices represent the end-points of each switch, and edges represent the connections between these end-points. When switch 1 is activated, an edge appears between the two end-points of switch 1.

Figure 2.8: Antenna Structure in [80].

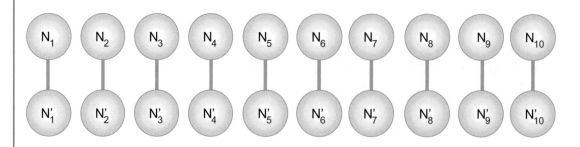

Figure 2.9: Graph Model of all possible configurations of the antenna in [80].

Rule 3: A multi-part antenna with parts connected by capacitors or varactors is modeled as a weighted undirected graph. The graph consists of a vertex for each antenna part. Undirected weighted edges are present wherever the parts have a physical connection.

<u>Constraints:</u>

The edges' weights in this case are calculated according to Eq. (2.5).

$$W_{ij} = P_{ij} + C_{ij\,normalized} \qquad C_{ij\,normalized} = \frac{C_{ij}}{C_{max}} \qquad (2.5)$$

$$\text{where} \quad P_{ij} = \begin{cases} 1 & A_{ij} = 0° \text{ or } 180° \\ 2 & \text{otherwise} \end{cases}$$

where A_{ij} represents the angle that the connection i, j form with the reference axis. C_{ij} represents the normalized capacitance of the capacitor connecting parts i and j, and C_{max} is the maximum capacitance incorporated on the antenna structure.

In this rule, P_{ij} is calculated as detailed in rule 1. The reference axis has to also be considered as explained in rule 1.

Example 2.4

As an example, let's take the antenna shown in Fig. 2.10 [81]. The antenna is a 2 × 2 reconfigurable planar wire grid antenna designed to operate in free space. Variable capacitors are placed at the centers of 11 of the 12 wire segments that comprise the grid. The values of the variable capacitors are constrained to lie between 0.1pF and 1pF. The graph modeling of this antenna follows rule 3 and is shown in Fig. 2.11. The vertices in this graph model represent the different parts of the lines that are connected together via a capacitor. The reference axis is taken along the vertical axis.

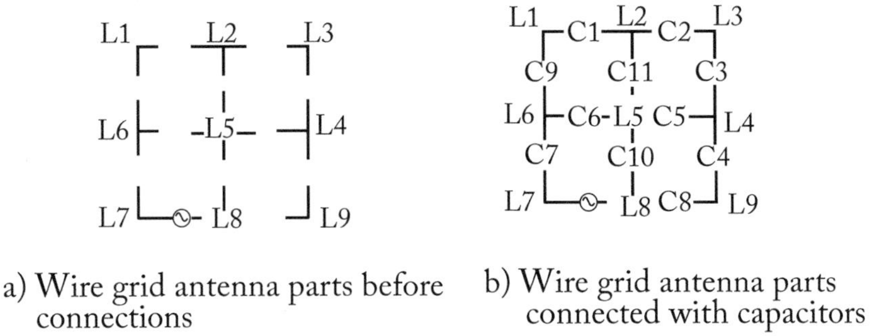

a) Wire grid antenna parts before connections

b) Wire grid antenna parts connected with capacitors

Figure 2.10: Antenna structure in [81].

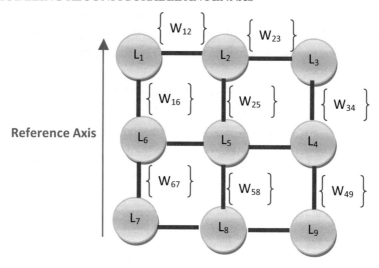

Figure 2.11: Graph model of the antenna in [81].

The values of the capacitors are not specified in [81]. The weights of each edge can be calculated according to Eq. (2.5) and as shown below in the adjacency matrix A.

$$A = \begin{bmatrix} 0 & W12 & 0 & 0 & 0 & W16 & 0 & 0 & 0 \\ W21 & 0 & W23 & 0 & W25 & 0 & 0 & 0 & 0 \\ 0 & W32 & 0 & W34 & 0 & 0 & 0 & 0 & 0 \\ 0 & 0 & W43 & 0 & W45 & 0 & 0 & 0 & W49 \\ 0 & W52 & 0 & W54 & 0 & W56 & 0 & W58 & 0 \\ W61 & 0 & 0 & 0 & W65 & 0 & W67 & 0 & 0 \\ 0 & 0 & 0 & 0 & 0 & W76 & 0 & 0 & 0 \\ 0 & 0 & 0 & 0 & W85 & 0 & 0 & 0 & W89 \\ 0 & 0 & 0 & W94 & 0 & 0 & 0 & W98 & 0 \end{bmatrix}$$

where

$$W_{12} = \frac{C1}{Max(C1,\ldots,C11)} + 2 \qquad W_{23} = \frac{C2}{Max(C1,\ldots,C11)} + 2$$

$$W_{34} = \frac{C2}{Max(C1,\ldots,C11)} + 1 \qquad W_{49} = \frac{C4}{Max(C1,\ldots,C11)} + 1$$

$$W_{45} = \frac{C5}{Max(C1,\ldots,C11)} + 2 \qquad W_{56} = \frac{C6}{Max(C1,\ldots,C11)} + 2$$

$$W_{98} = \frac{C8}{Max(C1,\ldots C11)} + 2 \qquad W_{85} = \frac{C10}{Max(C1,\ldots,C11)} + 1$$

$$W_{52} = \frac{C11}{Max(C1,\ldots,C11)} + 1 \qquad W_{76} = \frac{C7}{Max(C1,\ldots,C11)} + 1$$

$$W_{61} = \frac{C9}{Max(C1,\ldots,C11)} + 1$$

Rule 4: A single-part antenna with capacitors or varactors bridging over slots in its structure is modeled as a weighted undirected graph. This graph consists of a vertex for every capacitor or varactor's end-point. Weighted edges connect these vertices wherever capacitors or varactors are activated.

Constraints:

The graph should be undirected and weighted where the weights are defined in Eq. (2.6).

$$W_{ij} = C_{ij_{normalized}} \qquad (2.6)$$

where C_{ij} represents the normalized capacitance of the capacitor connecting end-points i and j. The capacitances values are calculated as discussed in rule 3. In the case of multiple slots, the same constraints imposed in rule 2 apply with the addition of Eq. (2.6).

Example 2.5

As an example, let's take the antenna shown in Fig. 2.12 [82]. Different varactors are used to connect different slots in the same radiating structure. Therefore, graph modeling of this antenna follows rule 4, where the vertices represent the end-points of the different varactors. The graph model consists of four vertices where each varactor corresponds to two vertices. The undirected edges are weighted with different varactor values. The graph model is shown in Fig. 2.13.

Rule 5: An antenna using angular or mechanical change in its structure to achieve reconfiguration is modeled with a weighted undirected graph. The vertices represent the different angular or mechanical positions of the physical action while edges between them represent the move from one position into the other.

Constraints:

The graph modeling of this type of antennas is undirected since the mechanical change (bending, rotation or, variable height) is reversible. The vertices represent the angles or mechanical positions of the corresponding physical action. The weighted edges connecting the vertices represent the change from one position to another. The weights represent the constraints associated with each change. For example, the time required to achieve a given rotation is considered as a constraint.

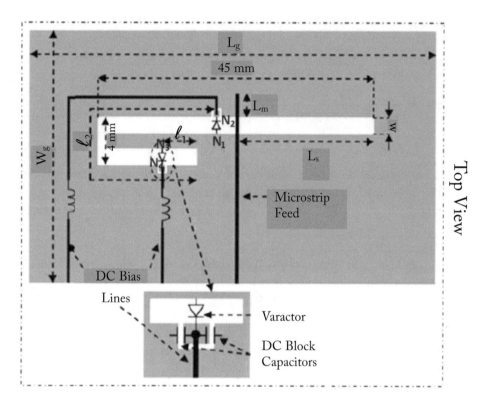

Figure 2.12: Antenna structure in [82].

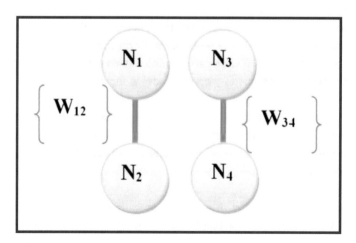

Figure 2.13: Graph modeling for the antenna in [82].

Example 2.6

As an example, let's take the antenna shown in Fig. 2.14 [83]. The antenna exhibits frequency tuning and a reconfigurable radiation pattern [83]. The graph modeling follows rule 5 where the bending angles are considered as vertices. The physical bending occurs as a response to an external field that is applied then removed when the antenna reaches a rest angle. The edges' weights which are the costs that a designer must pay may represent in this case the time of bending. The different weights can be evaluated as in Eq. (2.7):

$$w_{ij} = \begin{cases} 0 & i = j \\ T(A_i \rightarrow A_j) & i \neq j \end{cases} \qquad (2.7)$$

where $T(A_i \rightarrow A_j)$ represents the time it takes to bend the antenna from position i into position j. The adjacency matrix A shown below represents all the weights of the various edges. It can be evaluated numerically based on the fabricated system.

$$A = \begin{bmatrix} 0 & T(A_1 \rightarrow A_2) & T(A_1 \rightarrow A_3) & T(A_1 \rightarrow A_4) \\ T(A_2 \rightarrow A_1) & 0 & T(A_2 \rightarrow A_3) & T(A_2 \rightarrow A_4) \\ T(A_3 \rightarrow A_1) & T(A_3 \rightarrow A_2) & 0 & T(A_3 \rightarrow A_4) \\ T(A_4 \rightarrow A_1) & T(A_4 \rightarrow A_2) & T(A_4 \rightarrow A_3) & 0 \end{bmatrix}.$$

Figure 2.14: Antenna structure in [83].

Let's assume that the bending process takes into consideration four different positions as shown in Fig. 2.15(a). Therefore, the graph model must consist of four vertices. A_1 represents $0°$, A_2 represents $15°$, A_3 represents $45°$, and A_4 represents $90°$. The corresponding graph includes the four different positions:

$$Position\ 1 : 0° \quad A_1$$
$$Position\ 2 : 15° \quad A_1 \rightarrow A_2$$
$$Position\ 3 : 45° \quad A_1 \rightarrow A_2 \rightarrow A_3$$
$$Position\ 4 : 90° \quad A_1 \rightarrow A_2 \rightarrow A_3 \rightarrow A_4$$

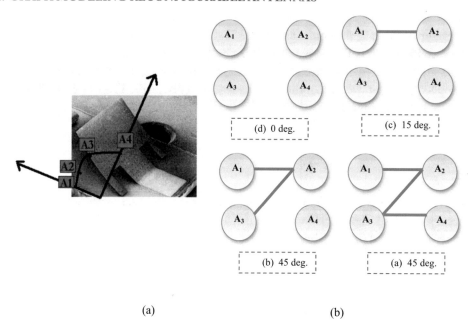

(a)   (b)

Figure 2.15: Graph modeling of the antenna in [83].

This is summarized in Fig. 2.16(b).

Rule 6: All the rules defined previously can be applied for the graph modeling of a reconfigurable feeding antenna. In this case, the reconfiguration is mainly achieved in the feeding network whether by switches, capacitors, or other means.

Constraints:

The graph components in this case represent the feeding components. All the rules' constraints defined previously apply correspondingly. For example, if the feeding reconfiguration is through switches then rule 1 or 2 applies and so on.

Example 2.7

As an example, let's take the antenna in [84]. This antenna is based on a parasitic antenna concept and it realizes pattern diversity. Each of the slot pairs in the parasitic patches is loaded by a switchable stub. The stub lengths are adjusted by p-i-n diode switches which allow four different patterns for each of the polarization states [84]. By switching ON a diode while the other one is OFF, the antenna can switch between horizontal or vertical polarization states with a single feeding port. Fig. 2.16 shows the feeding configuration connected by different switches (11 switches). The graph model according to rule 6 leads us to rule 1, where the vertices are the different lines in the feeding network connected together. The graph model is shown in Fig. 2.17 where the edges' weights are

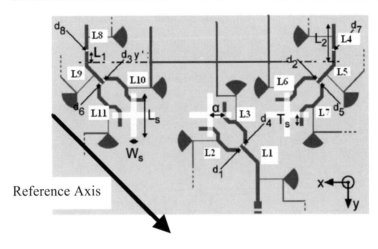

Figure 2.16: The feeding network of the antenna in [84].

calculated according to Eq. (2.4). The reference axis is taken as shown in Fig. 2.16. Four antenna states are considered in the graph model shown in Fig. 2.17.

Table 2.1 summarizes all the discussed rules with their various constraints and requirements.

2.4   APPLYING GRAPH ALGORITHMS ON RECONFIGURABLE ANTENNAS

The analysis of practical scenarios using graph models has led to the development of several techniques and methodologies that are designed to achieve optimal performance. Algorithms based on graphs have been heavily developed and optimized in many application areas such as control theory and networking. In this section, we highlight one algorithm called "Dijkstra's Algorithm" that illustrates a method that reveals the shortest path possible to get from point A to point B in any graph.

Dijkstra's algorithm [66] repeatedly selects the vertex with the minimum shortest-path estimate. For a given node in the graph, the algorithm finds the lowest cost path between that vertex and every other vertex. The algorithm can also be used to find the costs of shortest paths from one node to a certain destination node. The algorithm needs to stop once the shortest path has been identified. A summary of Dijkstra's algorithm operating procedure is shown in Fig. 2.18.

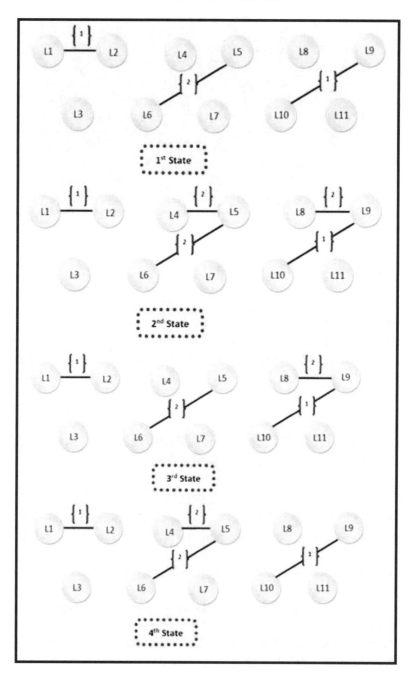

Figure 2.17: The graph model for the antenna in [84].

Table 2.1: A summary of the various rules for the graph modeling of reconfigurable antennas

	Multi-Part	Vertices	Directed	Weighted
Switches	YES	Parts	NO	YES
	NO	End-Points of the switches	NO	NO
Capacitors or varactors	YES	Parts	NO	YES
	NO	End-Points of the varactors	NO	YES
Mechanical Change	N/A	Positions	NO	YES
Reconfigurable Feeding	YES	Parts	NO	YES

2.4.1 APPLYING DIJKSTRA'S ALGORITHM TO THE CONTROL PROCESS OF RECONFIGURABLE ANTENNAS

In certain reconfigurable antennas a shorter path may mean a shorter current flow and thus a certain resonance associated with it. A longer path may denote a lower resonance frequency than the shorter path. In reconfigurable antennas resorting to physical and angular alterations, a shorter path means a faster response and a swifter reconfiguration.

The antenna in [85] resorts to slot rotation to achieve reconfiguration. Several commercial rotary switches can be used to automatically rotate the slot. Rotary switches can also be customized for this design and implemented through an FPGA (Field Programmable Gate Array) to control the rotation of the slots on the antenna.

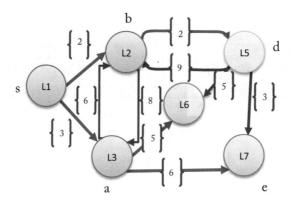

Figure 2.18: The shortest path shown in red as calculated by Dijkstra's algorithm for a weighted graph [66].

The graph model of this antenna follows rule 5 of Section 2.3. Vertices correspond to the antenna's different angles of rotation as shown in Fig. 2.19. The edges connecting these vertices are undirected and they represent the rotation process between two angles.

The cyclic flow of the graph is due to the fact that the graph is modeling the rotation process controlled by rotary switches. The mode of operation of rotary switches ensures a sequential rotation. For example, A1 represents 0 degree, A2 represents 30 degrees, and A3 represents 60 degrees. The rotation from 0 degree to 60 degrees is represented by edges connecting A1 (0 degree) to A2 (30 degrees) and A2 (30 degrees) to A3 (60 degrees).

The edges in the graph are weighted. In this model the weights represent the time of rotation from one angle into another and they are calculated according to Eq. (2.7). The graph modeling of this antenna with all possible edges is shown in Fig. 2.19.

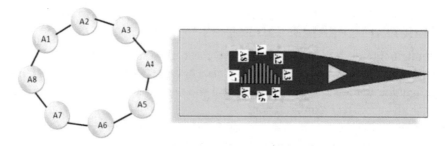

Figure 2.19: All possible configurations represented by all possible edges [85].

In the case where this antenna is implemented on a time sensitive platform like a satellite, arriving at the desired position with the shortest time possible, regardless of the system constraints, is important. Finding the shortest path for each possible scenario is a major requirement.

The complication of the rotational slotted antenna may vary from one system into another. For example, in a particular system, going from 0 to 30 degrees might be more costly than going from 0 to 230 degrees and thus the importance of application of algorithms like Dijkstra's algorithms. This algorithm can be used to program an FPGA to control the rotation of the slots through a rotary switch. The direction of rotation of the slots will be chosen according to the direction of the shortest path to move from one position into another.

2.5   DISCUSSION

In this chapter, rules for graph modeling various types of reconfigurable antennas are introduced. These rules are applied in the following two chapters (Chapters 3 and 4) to achieve new designs with a redundancy reduction technique. These rules are also used in Chapters 5, 6, and 7 to analyze the reliability and complexity issues of reconfigurable antenna systems.

CHAPTER 3

Reconfigurable Antenna Design Using Graph Models

3.1 INTRODUCTION

The design process of reconfigurable antennas lacks clear guidelines that set specific rules and constraints for optimal multifunctional antenna performance. Despite some publications where genetic algorithms are used as a design methodology for reconfigurable frequency selective surfaces [86], efforts in generating optimal reconfigurable antenna designs don't include any clear tactic.

Reconfigurable antenna structure redundancy has not been previously investigated very much. Presently, efficient designs are desired in order to reduce costs and losses. Usually, one proposes a design technique that satisfies given design constraints but without optimizing the number of redundant elements and minimizing unnecessary electronic components. Here, an iterative approach is introduced, in six steps, that can be applied to reconfigurable antenna structures, thus, eliminating any redundant element and achieving optimal structures. The structure redundancy methodology serves as a tool that is neither time consuming nor modeling exhaustive to iteratively tighten design constraints resulting in an optimal structure that delivers the required performance.

Graphs are used herein to model antenna structures with their reconfiguring components by using the rules detailed in Chapter 2. Antenna theory is joined in a common effort with graph models to produce non-redundant antenna structures that satisfy the design constraint set forth at initial design stages. In this chapter we discuss the proposed methodology in detail and present examples that verify the validity of the approach.

3.2 PROPOSED RECONFIGURABLE ANTENNA DESIGN STEPS

As a result of an observation process of various antenna design patterns in the literature, we propose six design steps that result in non-redundant reconfigurable antenna structures. These steps are proposed to be applied on new designs to be generated. The steps constitute a guideline for optimal antenna structure design.

Step 1: Specify the reconfigurability property that needs to be obtained.

First, the designer needs to define the reconfigurability property that is required from the antenna. These properties can be "frequency tuning, reconfigurable radiation pattern, reconfigurable polarization or different combinations." Based on the desired property the designer can later on, in the following steps, decide on the type of antenna to be designed, the reconfiguration technique to be employed, and fine tune the design accordingly.

Example on Design Step 1:

As an example we specify that the reconfigurability property needed here is: "frequency tuning and radiation pattern reconfigurability."

Step 2: Specify the antenna structure while considering the design constraints.

The designer at this step is required to have a general view of the structure of the antenna in question. The general shape and structure of the design is set in consideration with all the imposed constraints and without determining the exact dimensions and specifications.

Example on Design Step 2:

The designer is required to design a multi-band planar antenna (constraint 1). This antenna is also required to have five different configurations (constraint 2).

In order to meet these requirements the designer decides to propose intersecting microstrip lines that add multi-band operation to a main microstrip patch. Eight microstrip lines and a mid-section patch are required to satisfy the two constraints. Fig. 3.1(a) shows the corresponding structure [87].

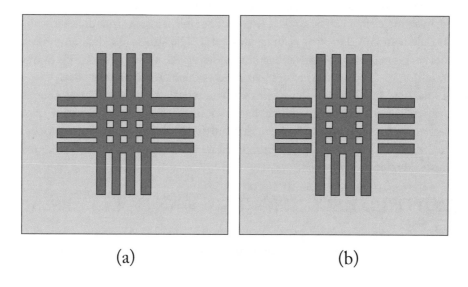

(a)   (b)

Figure 3.1: The proposed structure (a) Switches ON, (b) Switches OFF [87].

Step 3: Choose the reconfigurable technique.

The designer resorts to many techniques to design a reconfigurable antenna. Different categories of reconfigurable antennas require different design techniques. To design a frequency reconfigurable antenna a designer needs to alter the surface current distribution or reconfigure its feeding network accordingly. To produce an antenna with a reconfigurable radiation pattern, a designer needs to alter the radiating edges as well as reconfigure its feeding network. Polarization reconfigurable antennas are achieved by structure alteration. A reconfigurable antenna with multiple reconfiguration techniques requires the incorporation of various surface current alterations simultaneously. Several reconfiguration components can be used to accomplish these tasks such as switches, varactors, or mechanical techniques [88–106].

Example on Design Step 3:

Since the antenna in Fig. 3.1 needs to exhibit resonance tuning and a reconfigurable radiation pattern, then a combination of altering the surface currents and radiating edges apply. By using switches to connect and disconnect the mid-section from the microstrip lines shown in Fig. 3.1, the entire antenna structure changes leading to a modification in the surface currents distribution and the radiating edges. This yields the antenna shown in Fig. 3.1(b).

Step 4: Graph model the structure using proposed guidelines.

Graph model the design based on the general shape of the structure initiated in steps 2 and 3 by using the graph modeling rules and guidelines detailed in Chapter 2.

Example on Design Step 4:

Graph model the structure shown in Fig. 3.1(b). In this case, rule 1 is applied. We have a mid-section to which other parts are added. The vertices represent the various parts of the antenna structure. We call the vertex representing the mid-section P_0. The antenna is made of 9 parts; therefore the graph model includes 9 vertices. The different parts are added symmetrically and simultaneously as shown in Fig. 3.2. The edges connecting the vertices represent the connection of these parts to the mid-section by using electrical switches such as p-i-n diodes. The graph model is shown in Fig. 3.2.

Step 5: Fine tune the structure according to the desired applications using simulations and testing.

Simulate the structure outlined in step 2 and fine tune it to satisfy the existing constraints. At this point the design is defined and specified with exact dimensions. The designer is required to check with measurements, if necessary, the accuracy of the design and the fidelity of the response.

Example on Design Step 5:

Simulate the initial structure shown in Fig. 3.1 to satisfy the existing constraints.

The proposed antenna [87] consists of three layers. The lower layer is a square ground plane that covers the entire substrate and has a side length of 7 cm. The middle substrate has a dielectric constant $\epsilon_r = 3.9$ and a height of 0.16 cm. The upper layer is composed of eight microstrip lines intersecting with each other as shown in Fig. 3.3. The length of the microstrip lines are identical and optimized to correspond to $\lambda/2$ at 3.34 GHz (45 mm). The width of the microstrip line is taken as 3 mm to correspond to a characteristic impedance of 50 Ω. The optimized spacing between the lines

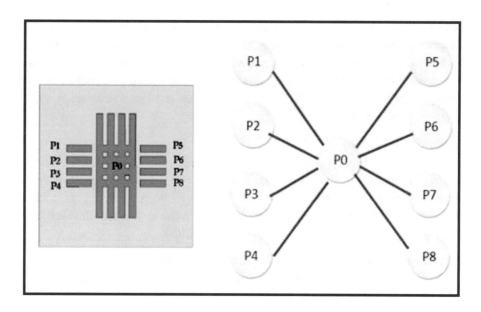

Figure 3.2: The graph model showing all possible connections.

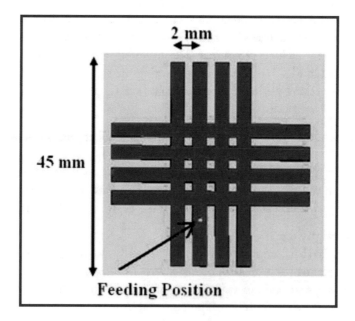

Figure 3.3: The proposed antenna structure at the initial stage.

is 2 mm. The choice of eight microstrip lines with optimal spacing intersecting with each other on the upper layer of the antenna is to achieve a multi-band antenna with a considerable bandwidth. The feeding position is optimized using simulations. The simulated and tested reflection coefficient results are compared in Fig. 3.4.

Achieving frequency and radiation pattern reconfiguration simultaneously requires a drastic change in the antenna structure through electrical means. Keeping the feeding position fixed, switches are used to connect different microstrip lines within the antenna structure. The reconfiguration of the antenna is achieved by using switches to control the length of the stubs. This allows the antenna to exhibit both frequency and radiation pattern reconfigurability. By referring to Fig. 3.2, the switches are used to connect P_0 to the different parts $P_1, P_2, \ldots, P_8$. Therefore, eight switches $\{S_1, \ldots, S_8\}$ are integrated within the antenna structure. As shown in Fig. 3.5 when the switches are off most of the resonances between 2.5 GHz and 3.5 GHz disappear and two new resonances appear between 2 GHz and 2.5 GHz. The radiation pattern is also affected by this action due to the change in the radiating elements as shown in Figs. 3.6 and 3.7. When the switches are activated, the maximum radiation in the E plane (Y-Z plane) is at 320° and for the H plane (X-Z plane) it is at 40°. The pattern totally changes when the switches are turned off.

Step 6: In case the designer is interested in finding an optimal solution for a configuration parameter such as redundancy. The designer should iteratively repeat steps 4 and 5 checking the feasibility of a solution as the constraints are tightened.

An iterative repetition of steps 4 and 5 with tightening the constraints results in a non-redundant structure. The designer needs to start by removing parts from the antenna structure while preserving the topological symmetry and characteristics. If by repeating steps 4 and 5 the designer ends up with the same antenna performance then the parts removed are redundant.

Example on Design Step 6:

An optimal antenna design has five configurations, which is interpreted physically by attaching two lines from each side to the mid-section. Five total parts preserve the symmetry of the structure and conserve the radiation pattern properties.

The graph model of the optimal antenna is shown in Fig. 3.8. The antenna is simulated with two parts from each side. The optimization of the 2 parts with the simulator has led to lines of width 0.9 cm and length 1.15 cm from each side of the mid-section. A comparison is made between the optimal antenna achieved in step 6 and the old redundant antenna by comparing the reflection coefficient as shown in Fig. 3.9 which proves that the parts removed are redundant and four switches are spared. The radiation patterns of the non-optimal and the optimal antennas at 5.5 GHz are compared in Fig. 3.10 in the Y-Z plane when the switches are not activated (OFF) which proves that the removal of the redundant parts and the redesigning of the structure did not affect the radiation pattern properties. The optimal antenna is now fabricated and tested and great analogy is shown between the tested and simulated S11 results as shown in Fig. 3.11. The optimized fabricated antenna and the original fabricated redundant antenna are shown in Fig. 3.12 for comparison.

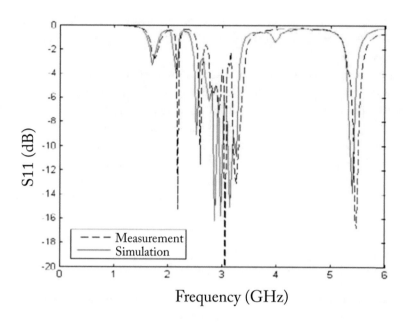

Figure 3.4: A comparison between the measured and simulated return loss [87].

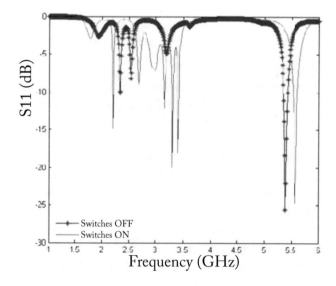

Figure 3.5: S11 resonance tuning for different configurations [87].

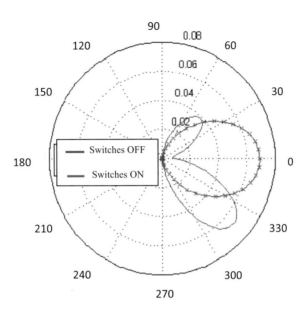

Figure 3.6: E plane (Y-Z Plane) radiation pattern reconfigurability at $F = 2.33$ GHz [87].

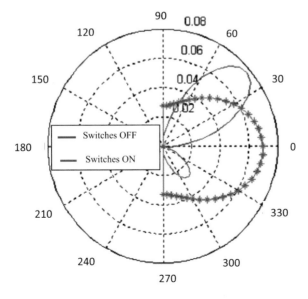

Figure 3.7: H plane (X-Z) radiation pattern reconfigurability at $F = 2.33$ GHz [87].

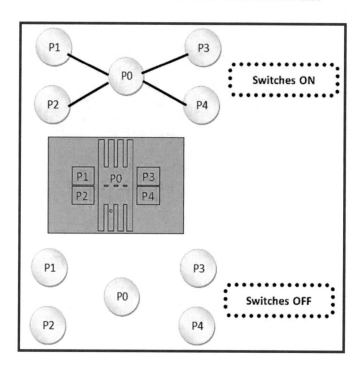

Figure 3.8: The graph model of the iteratively optimized structure [47].

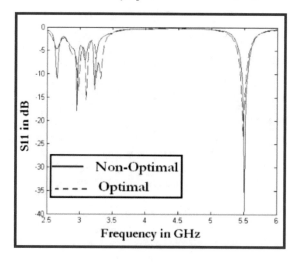

Figure 3.9: Comparison between the S11 results for the non-optimal and the optimal antenna when the switches are activated [47].

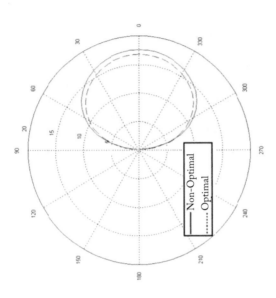

Figure 3.10: The (Y-Z) plane radiation pattern at $F = 5.5$ GHz for the non-optimal and the optimal antenna when the switches are open [47].

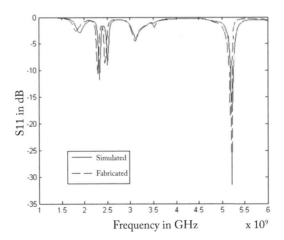

Figure 3.11: A comparison between the simulated and tested S11 results for the optimal antenna [47].

Optimal Antenna   Non-Optimal Antenna

Figure 3.12: The fabricated optimized and original antennas [47].

3.2.1   SUMMARY OF THE PROPOSED DESIGN TECHNIQUE

The design technique is summarized in the chart in Fig. 3.13. This chart details the progression and the interrelation of the various steps. The chart can be transformed into an algorithm that can be used to automate the iterative optimization process.

3.3   EXAMPLE OF DESIGNING A RECONFIGURABLE ANTENNA USING THE PROPOSED ITERATIVE DESIGN STEPS

In this section we take an example of a previously published antenna [98] and we examine its optimal topology. The iterative design approach is applied on the antenna examined and the constraints are tightened to realize an optimal topology. The antenna design is shown in Fig. 3.14 [98]. This antenna resonates, first at 1.85 GHz and 3.2 GHz, then at 1.85 GHz and 3.4 GHz, and finally at 1.85 GHz and 3.6 GHz. This antenna tunes the lower frequency from 1.85 GHz to 2.4 GHz. The higher frequencies remain unchanged. Four symmetrical parts are added for each length [98]. In this section the antenna is redesigned and the decision to add four symmetrical parts is re-examined. Applying the various steps of Section 3.2 results in the following:

Step 1: The reconfigurability property is a tunable reflection coefficient.

Step 2: The antenna required is a microtrip antenna. The material used is Rogers RO 3203 with 1.524 mm of height. The shape of the patch is a monopole, and having a coplanar ground plane is one of the constraints.

Step 3: Since this antenna is required to exhibit a tunable reflection coefficient; then a surface current distribution or a change in the feeding network is required. Switches are chosen to connect different parts to the monopole which satisfies a second design constraint.

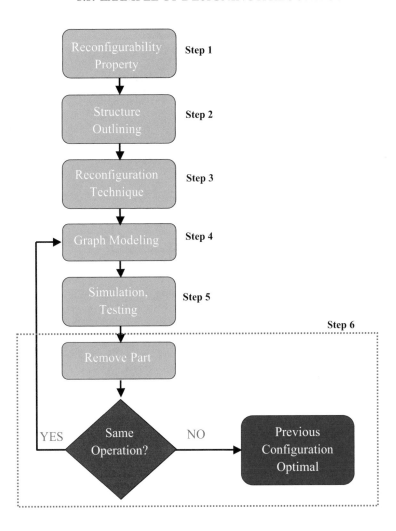

Figure 3.13: A chart representation of the proposed design technique.

Step 4: Graph model the structure using the rules detailed in Chapter 2. For this antenna rule 1 applies. The antenna consists of a main monopole to which other parts are added. The vertices of the graph represent the different parts. The vertex representing the main monopole is named "P_0" and the added monopole "P_1." Three frequency changes are required for each length of the monopole; this means three antenna configurations are required for each monopole length. The four symmetrical parts added in [98] are represented by P_2, P_3, P_4, and P_5. The edges between the vertices represent the connection of the different parts. These edges are weighted and the weights are

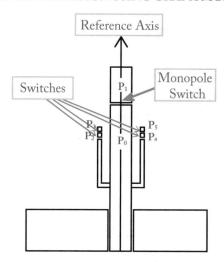

Figure 3.14: Antenna structure in [98] with parts numbered.

calculated according to Eq. (2.4). The main monopole's direction is considered as the reference axis as shown in Fig. 3.14. The graph modeling of this antenna with all possible connections is shown in Fig. 3.15. The adjacency matrix A representing all the weights for the antenna with a total of six parts is shown below.

$$A = \begin{bmatrix} W_{00} & W_{01} & W_{02} & W_{03} & W_{04} & W_{05} \\ W_{10} & W_{11} & W_{12} & W_{13} & W_{14} & W_{15} \\ W_{20} & W_{21} & W_{22} & W_{23} & W_{24} & W_{25} \\ W_{30} & W_{31} & W_{32} & W_{33} & W_{34} & W_{35} \\ W_{40} & W_{41} & W_{42} & W_{43} & W_{44} & W_{45} \\ W_{50} & W_{51} & W_{52} & W_{53} & W_{54} & W_{55} \end{bmatrix} = \begin{bmatrix} 0 & 1 & 0 & 2 & 2 & 0 \\ 1 & 0 & 0 & 0 & 0 & 0 \\ 0 & 0 & 0 & 1 & 0 & 0 \\ 2 & 0 & 1 & 0 & 0 & 0 \\ 2 & 0 & 0 & 0 & 0 & 1 \\ 0 & 0 & 0 & 0 & 1 & 0 \end{bmatrix}$$

Step 5: The design is simulated. The dimensions of the parts are optimized so that if the monopole switch shown in Fig. 3.14 is off, 1.85 GHz is obtained and if the switch is on, 2.4 GHz is obtained.

Step 6: The designer's decision is not optimal since comparing between the number of antenna configurations and the number of resonances obtained; we find that two parts can be easily omitted. An iterative repetition of steps 4 and 5 would result in removing the antenna redundancies. The desired optimal structure is graph modeled in Fig. 3.16. The same functional behavior should be obtained from the optimized antenna. The optimized antenna structure has four parts instead of six and two switches are eliminated. The optimized antenna resembles the original one in Fig. 3.14, except that it has only four total parts instead of six. The adjacency matrix of the graph representing the optimized structure is shown below:

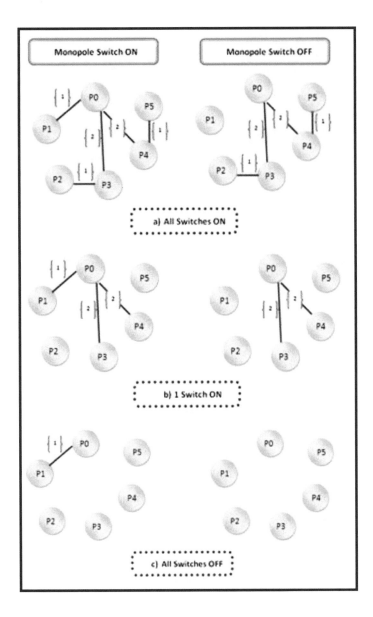

Figure 3.15: Graph modeling before designing the antenna showing the different antenna configurations.

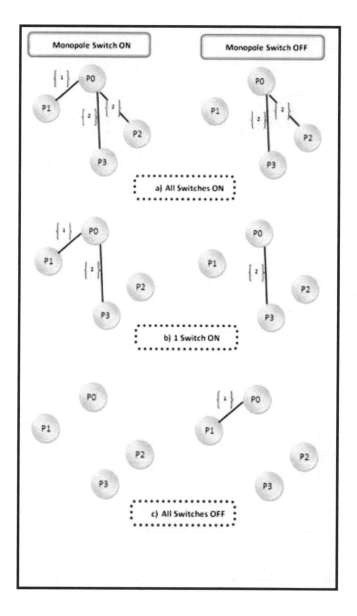

Figure 3.16: Graph modeling for the optimal antenna design.

$$A_{opt} = \begin{bmatrix} 0 & 1 & 2 & 2 \\ 1 & 0 & 0 & 0 \\ 2 & 0 & 0 & 0 \\ 2 & 0 & 0 & 0 \end{bmatrix}$$

3.4   DISCUSSION

In this chapter we have presented a new design technique based on an iterative optimization approach. The technique asks the user to keep tightening the design constraints while maintaining the same performance. As a result an optimal structure is achieved.

This iterative approach is divided into six steps and is applied on two antenna examples. The design approach can be software executed. This approach applied to large antenna structures may result in a time consuming process. Thus, in the next chapter a redundancy reduction approach is introduced and formulated for any type of reconfigurable antennas. The main purpose is to present the reconfigurable antenna designers with easy tools for efficient antenna design processes.

CHAPTER 4

Redundancy Reduction in Reconfigurable Antenna Structures

4.1 INTRODUCTION

In this chapter we provide the designer with a new and easy tool to design a reconfigurable antenna with only the necessary number of elements to perform the required functions. The designer is asked to tailor an antenna design that only satisfies the required constraints without any disturbance of the antenna's radiation characteristics. The designer will apply this technique to a new or previously designed reconfigurable antenna to remove redundant elements from its structure. This technique provides the designer with a clear guideline to generate constraint satisfying designs with performance limited to the required functions. The technique eliminates additional unnecessary elements from the antenna structure and generates an optimal topology using equations and formulations that are derived from graph models and antenna theory. This redundancy removal approach presents an easier and faster method than the iterative approach presented in the previous chapter. It is a structure optimization process that results in a design that only achieves the required functions. The functions required and the constraints imposed on the designer by the system are taken into consideration before the elimination of any redundant element.

Various optimization algorithms have been proposed in the literature to be incorporated with a reconfigurable antenna platform. However, these optimization algorithms are used to smooth the state transition in a reconfigurable antenna's operation. In [86] the flexibility of a grid of metallic patches interconnected by a matrix of switches is exploited by optimizing the switch settings using a genetic algorithm to produce a desired frequency response. Genetic algorithms are also used in [107] to search the patterns that the antenna might produce for a given frequency.

Neural networks are used to select which switching device should be activated for each reconfiguration state [89]. Each group of selected switches allows the antenna to operate within a certain band. This task is handled as a classification type of problem and is accomplished by a self-organizing map neural network (SOM NN) [108].

A method based on the clonal selection algorithm (CLONALG) is used to design a reconfigurable dual-beam linear antenna array with excitation distributions differing only in phase [109]. CLONALG is a relatively novel population-based evolutionary algorithm inspired by the clonal selection principle of the human immune system.

In order to decide which optimization method is the most convenient for a specific design, different algorithms are compared together [110]. The use of genetic algorithms, simulated annealing, and ant-colony optimization applied to reconfigurable antennas is investigated in [110]. All these algorithms are compared to the random search method. The work shows that each optimization algorithm outperforms the random search method [110]. A comparison between genetic algorithms and particle swarm optimization, versus the cross-entropy method is shown in [111]. Results show that the cross-entropy method has a fast convergence speed, but it needs large population size to function [111]. A particular optimization algorithm cannot be separated from the rest as the best fit before selecting a specific reconfigurable scheme [112].

In this chapter we present a new redundancy reduction approach. This approach constitutes a structure optimization technique. This technique resorts to the graph modeling rules previously defined in Chapter 2. Based on these rules a set of equations is formulated. These equations indicate the presence or lack of redundancy in an antenna structure and in the number of switching elements. This approach proposes a way of reducing the antenna structure's complexity and thus can't be compared to the optimization techniques discussed previously such as genetic algorithms, clonal selection, or ant colony, due to the difference in objective. It is not a function optimization technique but rather a structure optimization methodology.

4.2   ANTENNA STRUCTURE REDUNDANCY REDUCTION

A part is defined as redundant if its presence gives the antenna more functions than required and its removal does not affect the antenna's desired performance [113]. The removal of a part from the antenna structure may require a change in the dimensions of the remaining parts in order to preserve the antenna's original characteristics, i.e., a redundant part can be removed as long as its removal will not affect the polarization status of the antenna that exhibits frequency tuning and polarization diversity [113].

Every path in every graph model representing the reconfigurable antenna should correspond to a different configuration. If the number of unique paths (NUP) in the graph model is larger than the number of configurations, then redundancy may exist in the antenna structure. An example of counting the total number of unique paths in a graph model is shown in Fig. 4.1 [113]. The graph containing all possible connections shows the presence of four vertices with three undirected edges. Thus, a total of six unique paths exist in this graph ($NUP = 6$). These unique paths are shown each individually in Fig. 4.1. It is important to note that since the edges shown in the graph are undirected, the paths connecting any two vertices are bidirectional. The paths are as follows:

- Path connecting P_1 to P_3, or P_3 to P_1

- Path connecting P_1 to P_2, or P_2 to P_1

- Path connecting P_3 to P_2, or P_2 to P_3

- Path connecting P_1 to P_0, or P_0 to P_1

- Path connecting P_0 to P_3, or P_3 to P_0

- Path connecting P_0 to P_2, or P_2 to P_0

If redundant parts were removed from the antenna structure their corresponding vertices and edges should also be removed from the graph model.

Since each unique path represents a unique antenna function, it is important to identify the number of possible unique paths. In our redundancy reduction approach, only reconfigurable antennas using one reconfiguration technique are investigated. If an antenna uses more than one reconfiguration technique then each technique is investigated separately.

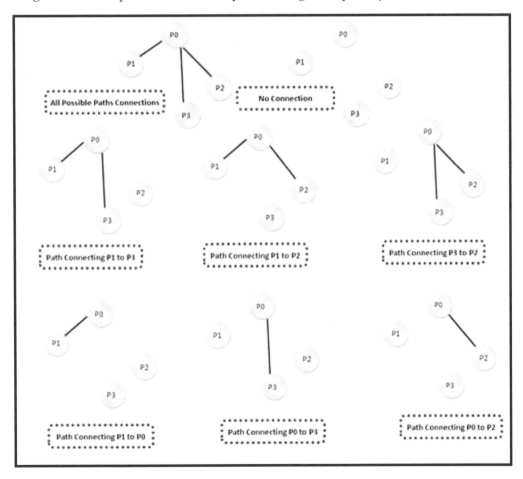

Figure 4.1: An example of possible unique paths in a given graph [113].

4.2.1   THE TOTAL NUMBER OF EDGES IN A COMPLETE GRAPH

A complete graph is a graph in which each vertex is connected by an edge to each of the other vertices. Suppose the set of vertices in a complete graph model is $V = \{1, 2, \ldots, N\}$. A vertex i can be selected in n ways, i.e., there are exactly $(N - 1)$ edges between a selected vertex i and the remaining $(N - 1)$ vertices [72–77, 114]. In Fig. 4.2 a complete graph with three vertices is shown, where each pair of vertices is interconnected. Each vertex has $(N - 1) = 2$ edges with the remaining two vertices as shown in Fig. 4.2.

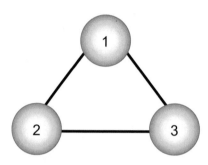

Figure 4.2: A complete graph with three vertices.

The edge joining vertex i with another vertex j is the same as the edge joining j and i. Thus, the total number of edges K_N in the complete graph is:

$$K_N = \frac{N(N - 1)}{2} \tag{4.1}$$

where:

$$N = \text{number of vertices.}$$

The total number of edges in the graph shown in Fig. 4.2 is: $K_3 = \frac{3(3-1)}{2} = 3$ edges.

The total number of edges in the complete graph composed of six vertices and shown in Fig. 4.3 is:

$$K_7 = \frac{6(6 - 1)}{2} = \frac{6 \times 5}{2} = 15$$

4.2.2   DERIVING EQUATIONS FOR REDUNDANCY REDUCTION IN MULTI-PART ANTENNAS

In order to set some equations for reducing redundancy in a multi-part antenna structure, and based on the fact that each unique path corresponds to a unique function, the number of unique paths in a graph has to be minimized.

Formulating the total number of unique paths in any graph model has not been solved yet. According to the graph modeling rules defined in Chapter 2 for multi-part antennas, graphs representing these antennas are arbitrary and depend on the antenna topology which is not generalized

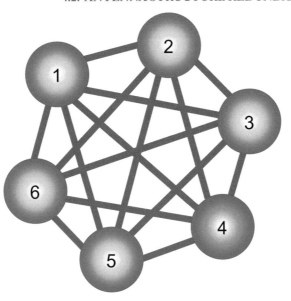

Figure 4.3: A complete graph with six vertices.

and may be considered random. Thus, the number of unique paths cannot be formulated and minimum bound needs to be estimated in order to minimize the number of RF components (switches) used and reduce redundancies.

According to the rules of Chapter 2, the graph model of a multi-part antenna assigns a vertex for each part that is connected to another part through an edge. Thus, for N vertices in a graph representing a multi-part antenna, a minimum of $(N - 1)$ edges exist. There can't be less than $(N - 1)$ edges or else idle vertices representing unnecessary elements would exist. The maximum number of edges is equal to K_n from Eq. (4.1), which is the total number of edges in a complete graph. Thus, the number of edges NE in a graph model representing a multi-part antenna is bounded by the following:

$$(N - 1) \leq NE \leq \frac{N(N - 1)}{2} \tag{4.2}$$

For the minimum number of edges $(N - 1)$, the number of existing unique paths (NUP) in a graph model representing a multi-part antenna is always greater than $K_N = \frac{N(N-1)}{2}$ or else idle vertices are present in that graph model.

In order to minimize the number of unique paths, the necessary number of unique paths required in a multi-part antenna graph model is taken to be equal to $\frac{N(N-1)}{2}$. By decreasing the number of unique paths, the number of possible antenna configurations is reduced which results in reducing the number of vertices and redundant parts are removed. Therefore, NUP is taken as:

$$NUP = \frac{N(N-1)}{2} \qquad (4.3a)$$

The number of available antenna configurations (NAC) is equal to the number of unique paths in addition to the configuration where no edge exists. Different antenna configurations are related to different unique paths in addition to the case where all switches are off. Thus, NAC can be formulated as in Eq. (4.3b):

$$NAC = NUP + 1 \qquad (4.3b)$$

Therefore, the minimum number of vertices can be derived from (4.3a) and (4.3b) as in Equation (4.3c):

$$N^2 - N - 2 \times (NAC - 1) = 0 \Rightarrow \quad N = \left\lceil \frac{1 + \sqrt{1 + 8 \times (NAC - 1)}}{2} \right\rceil \qquad (4.3c)$$

The reconfigurable antenna may have more possible configurations than NAC for a given set of vertices, however NAC represents the minimum bound of configurations that are necessary to achieve a constraint satisfying design.

4.2.3   DERIVING EQUATIONS FOR REDUNDANCY REDUCTION IN SINGLE-PART ANTENNAS

In the case of single-part antennas, each vertex in a graph represents an end-point of an RF switch bridging over a slot and each edge represents the connection between these two end-points whenever the corresponding switch is activated. Thus, the number of vertices N is double the number of all possible edges. In this case, the number of unique paths (NUP) is equal to the number of possible edges in the graph. The minimum number of available antenna configurations (NAC), to achieve a constraint satisfying design can be defined as the number of possible edges in addition to the case where no connection exists.

$$NAC = \frac{N}{2} + 1 \qquad (4.4a)$$

The reconfigurable antenna may have more possible configurations than NAC for a given set of vertices since NAC only represents the minimum bound. The number of vertices necessary to achieve a certain number of configurations is:

$$N = 2 \times (NAC - 1) \qquad (4.4b)$$

4.2.4   DERIVING EQUATIONS FOR REDUNDANCY REDUCTION IN ANTENNAS RESORTING TO MECHANICAL RECONFIGURATION METHODS

The graph model in this case represents reconfigurable antennas using angular or mechanical change as a reconfiguration technique. For each physical position or angle a different configuration is possible.

Thus, the number of available configurations (NAC) is equal to the number of vertices in the graph

$$NAC = N \tag{4.5}$$

4.2.5   A CHART REPRESENTATION OF THE REDUNDANCY REDUCTION APPROACH

As a summary of the redundancy reduction approach we can represent the whole process in a chart as shown in Fig. 4.4 [47]. The process debuts with the graph modeling of reconfigurable antenna structures based on the rules of Chapter 2. The equations derived in Section 4.2 are then applied to the corresponding graph model. A simple comparison between the resulting *NAC* and the number of required functions determines the presence or lack of redundant elements. If redundant elements exist, then the designer has to decide whether or not the topology of the antenna being optimized needs to be preserved or not. If yes, then a reduction of the RF reconfiguring components is achieved; if the topology can be altered then the whole antenna needs to be redesigned with non-redundant elements.

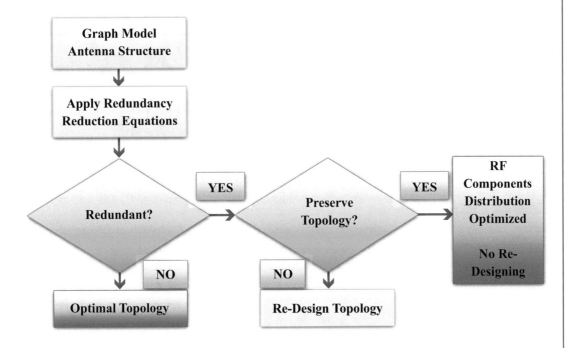

Figure 4.4: A chart representation of the redundancy reduction approach.

4.3   EXAMPLES

In this section we apply the derived redundancy reduction equations on antenna examples to prove their validity. The antenna examples are tested for redundancy. If the antennas are found to be redundant, the redundant elements are identified and removed. Non-redundant structures are then proposed and redesigned.

Example 4.1

In this section, the antenna discussed in the previous chapter is considered. The antenna is required to have resonance tuning and radiation pattern reconfigurability. The design as well as the graph model is presented in Fig. 4.5. This antenna is required to have five different configurations. Since this is a multi-part switch-reconfigured antenna, NUP and NAC are:

$$NUP = \frac{N(N-1)}{2} = \frac{9(9-1)}{2} = 36$$

$$NAC = NUP + 1 = 37$$

This antenna has a minimum of 37 possible configurations.

Since the total number of possible configurations is larger than the required antenna configurations, redundancy exists. In order for the antenna to present only five configurations without compromising its desired performance, NAC should be equal to 5. Therefore, the required number of vertices in the graph model is calculated as:

$$N = \left\lceil \frac{1 + \sqrt{1 + 8 \times (NAC - 1)}}{2} \right\rceil = \left\lceil \frac{1 + \sqrt{1 + 8 \times (5 - 1)}}{2} \right\rceil = 4$$

To achieve a non-redundant design, the number of vertices should be reduced to 4. The graph model with four vertices is shown in Fig. 4.6(a). It is composed of three vertices (P_1, P_2, P_3) connected to a main vertex (P_0). This graph model corresponds to an antenna with three parts attached to a main part. It is important to note that this graph model doesn't preserve the structure's symmetry. The optimization technique allows the removal of redundant parts as long as their removal does not affect the antenna characteristics such as symmetry. Therefore, four total parts as represented by the four vertices, is not a good solution for this antenna and in such a case, $N > 4$ is required. Taking $N = 5$ preserves the antenna's symmetry and gives the antenna an $NAC = 11$. The corresponding graph is shown in Fig. 4.6(b).

$$NUP = \frac{N(N-1)}{2} = \frac{5(5-1)}{2} = 10$$

$$NAC = NUP + 1 = 11$$

Satisfying all the required constraints and functions as described in Chapter 3, the graph can be translated into four different sections connected to a main antenna patch. The dimensions and

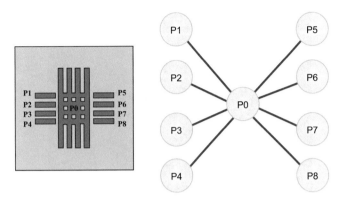

Figure 4.5: Antenna structure in [87] and its graph model.

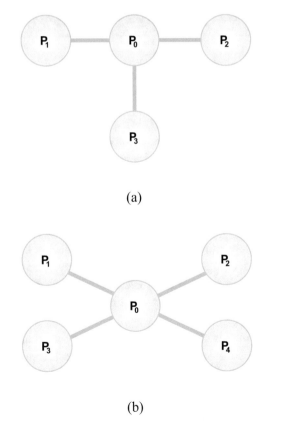

(a)

(b)

Figure 4.6: (a) Graph model with four vertices, (b) Graph model with five vertices.

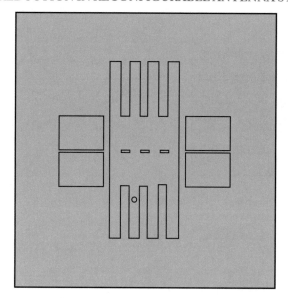

Figure 4.7: The optimized antenna structure when the switches are all ON or all OFF.

size of each of the patches can be determined using a simulator. The resulting antenna topology is shown in Fig. 4.7 and is seen to preserve the symmetry of the structure. To verify the validity of this approach, the original and the optimized antennas are simulated. This approach matches the iterative optimization procedure of Section 3.2. The similarity between the antenna's radiation patterns confirms that the removal of the redundant parts does not drastically affect the radiation characteristics as shown in Fig. 3.10. The different reflection coefficient results for four different configurations with actual p-i-n diodes installed on the non-redundant antenna is shown in Fig. 4.8.

The method discussed here and the iterative design approaches discussed in Chapter 3 have resulted in the same results. In fact, these two techniques are complimentary. The iterative approach of Chapter 3 allows the removal of a part at every cycle of design, and the constraints are tightened until the desired function is compromised. Once this limit is reached the optimal design with non-redundant element is achieved.

The equations derived in this chapter are based on the fact that each unique path should indicate a unique function, thus the removal of a unique path will result in the loss of a function. This is an indication of tight constraints that have no redundancy tolerance. Therefore, the equations governing the redundancy reduction approach in this chapter are a direct derivation of this concept where the limits for number of unique paths (NUP) are chosen to only satisfy required functions and nothing beyond that.

If one asks a question, about which method to use and when, the answer should indicate that both methods reveal the same results. The size of the reconfigurable antenna structure as well as its

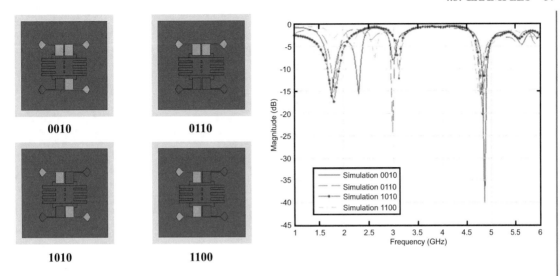

0010   **0110**

1010   **1100**

Figure 4.8: Reflection coefficient for four different configurations of the optimized antenna [11]. The activated parts are shown in blue on the left.

required functions may constitute an indication; however the approach in this chapter is designed to provide a faster and simpler redundancy reduction approach that will generate non-redundant designs based on graph models topologies.

Example 4.2

As an example, we take the switch-reconfigured antenna shown in Fig. 4.9 [101]. This antenna is built out of a hexagonal main patch and six trapezoidal parts placed around it. The graph model of this antenna conforms to rule 1 (multi-part antenna). This antenna is required to have the following four configurations:

Configuration 1: 1 GHz, 3.5 GHz, 4.5 GHz

Configuration 2: 3.5 GHz, 4.5 GHz, 5 GHz

Configuration 3: 1 GHz, 2.5 GHz, 5 GHz

Configuration 4: 3 GHz, 3.5 GHz, 4.5 GHz

These frequencies represent practical applications such as WiMAX, WiFi, and GPS. This antenna is originally designed in [101] to have six switches connecting six sections to a main section. The graph model of this antenna shown in Fig. 4.9(b), indicates that its corresponding $NAC = 22$:

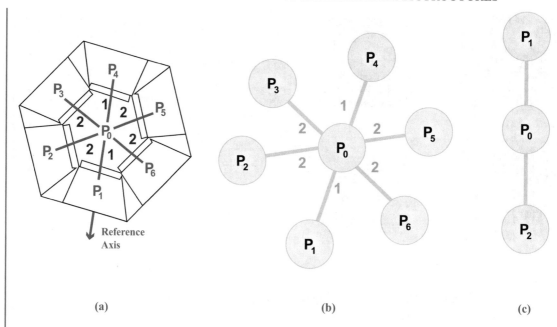

(a)   (b)   (c)

Figure 4.9: (a) Antenna in [101], (b) Graph model of the original antenna, (c) Graph model of the non-redundant antenna.

$$NUP = \frac{N(N-1)}{2} = \frac{7(7-1)}{2} = 21$$

$$NAC = NUP + 1 = 22$$

While the required configurations are only 4. Therefore, one concludes that redundant parts exist and they must be removed as long as the antenna's radiation characteristics are preserved.

To achieve these four required configurations, ($NAC = 4$), the number of vertices in the graph representing the non-redundant antenna is:

$$N = \left\lceil \frac{1 + \sqrt{1 + 8 \times (NAC - 1)}}{2} \right\rceil = \left\lceil \frac{1 + \sqrt{1 + 8 \times (4 - 1)}}{2} \right\rceil = 3$$

The corresponding graph model of the non-redundant antenna is shown in Fig. 4.9(c).

To reduce the redundancy of this system and to minimize the design time and the number of simulations, the number of switches used has to be reduced to two. The designer optimizes by simulations the placement of the two switches to achieve the required frequencies and configurations.

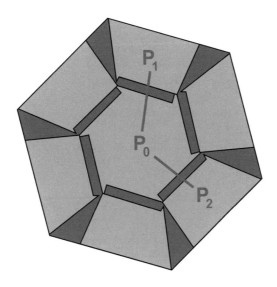

Figure 4.10: The optimized structure with its graph model [113].

However, to preserve the radiation properties, the general shape of the antenna as a six-armed hexagon cannot be disturbed, especially when all switches are OFF [113]. The placements of these switches on the optimized antenna are shown in Fig. 4.10 [113]. The simulated reflection coefficient for all required configurations is shown in Fig. 4.11. By applying this technique, the design time has been reduced and, instead of determining the placement and topology of the antenna with six switches, the work has to be done for only two switches. A comparison of the antenna's radiation patterns with redundant switches and the one without redundant elements at 4.517 GHz is shown in Fig. 4.12 for the x-y and y-z plane cuts.

Example 4.3

In this example the antenna shown in Fig. 4.13(a) [115] is studied. It is a MEMS-reconfigurable pixel structure that provides two functionalities: reconfiguration of its modes of radiation and reconfiguration of the operating frequency. The proposed antenna uses a 13 × 13 matrix of metallic pixels interconnected through MEMS switches in which circular patches of different radii are mapped. Each metallic pixel has dimensions of 1.2 × 1.2 mm and they are separated by 2 mm, to provide enough space to allocate the MEMS switches and connecting biasing lines. The MEMS switches around each pixel are activated or deactivated depending on the DC voltage that is supplied to these pixels. The DC connectivity is provided through bias lines that connect the pixels to the back side of the substrate, as shown in Fig. 4.13(b).

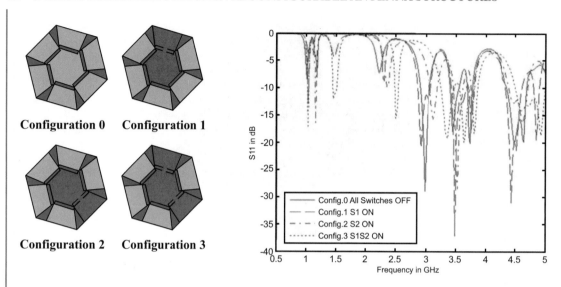

Figure 4.11: The reflection coefficient plot for required configurations. The activated parts are shown in red on the left [113].

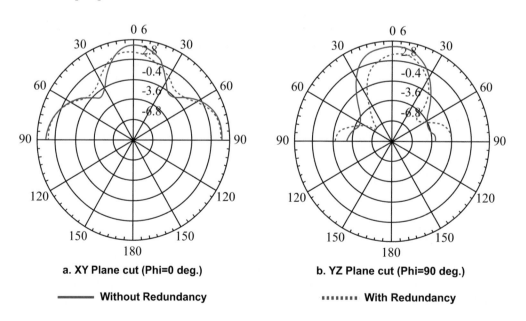

Figure 4.12: Comparison of the antenna's simulated radiation patterns with and without redundant elements at 4.517 GHz for the x-y and y-z plane cuts [113].

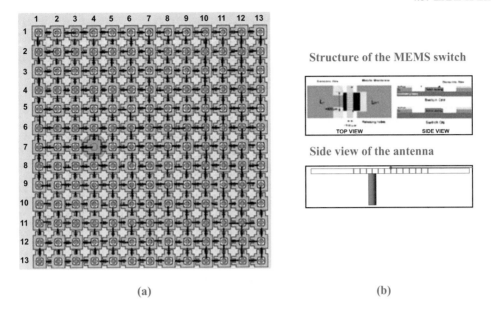

Figure 4.13: (a) Structure of the multifunctional MEMS-reconfigurable pixel antenna. (b) Structure of the MEMS switch as well as a side view of the antenna [115].

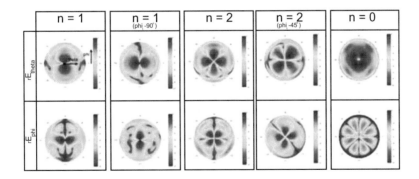

Figure 4.14: Flattened 3-D radiation pattern with respective antenna configurations [115].

This antenna can generate five orthogonal radiation patterns at any frequency between 6 and 7 GHz. Thus, the corresponding antenna exhibits the following modes:

Mode 1: $n = 1$, $\Phi_0 = 0°$; Mode 2: $n = 2$, $\Phi_0 = 0°$; Mode 3: $n = 0$, $\Phi_0 = 0°$; Mode 4: $n = 1$, $\Phi_0 = 90°$; Mode 5: $n = 2$, $\Phi_0 = 45°$ [116]. These modes of operation are shown in Fig. 4.14. At any fixed frequency between 6 and 7 GHz, five radiation states can be selected. Fig. 4.13(a) shows the antenna with all possible connections.

Fig. 4.15 shows the different configurations required from the antenna to achieve the different radiation pattern characteristics.

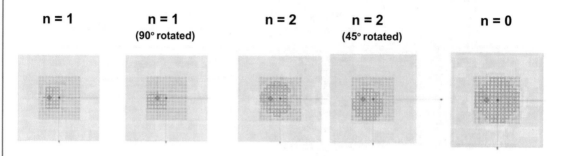

n = 1   **n = 1**   **n = 2**   **n = 2**   **n = 0**
(90° rotated)   (45° rotated)

Figure 4.15: Different configurations required from the antenna [115].

To graph model this antenna the parts constituting its structure are considered as vertices. These vertices are connected by weighted undirected edges. The graph model of the different antenna configurations required to achieve the five different modes of operation is shown in Fig. 4.16 and follows rule 1 of graph modeling.

Since only five configurations are required then the corresponding number of parts required to achieve the desired configurations should be:

$$N = \left\lceil \frac{1 + \sqrt{1 + 8 \times (NAC - 1)}}{2} \right\rceil = \left\lceil \frac{1 + \sqrt{1 + 8 \times (5 - 1)}}{2} \right\rceil = 4$$

Only four configurations are required to achieve five antenna functions. The shape of the antenna with four parts will be very different from the one shown in Fig. 4.13 and needs to be simulated and investigated extensively. The antenna designed in [115] required a minimization of the number of switches used while keeping the same antenna topology. To preserve the same antenna topology, redundant connections have to be identified and redundant switches need to be eliminated. By comparing the different graph models of Fig. 4.16 one notices that edges connect only certain vertices and the rest of the vertices remain idle in all five configurations.

To determine which switches to remove and identify the redundant elements, a new analysis approach is applied here. The size of the antenna with the large number of switches implies that the graph models representing the different states of the antenna topology have large sizes (number of edges). It may be easier to express the graphs as adjacency matrices and then determine the intersection between each adjacency matrix. The intersection between various adjacency matrices reveals the required connections as well as identifies idle elements. For simplification purposes we have decided to consider the edges to be non-weighted. Thus, a part connected by an edge has a

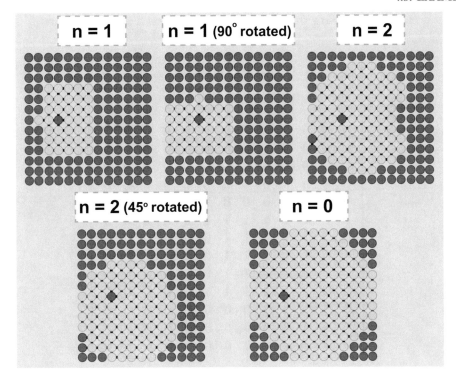

Figure 4.16: The graph model of the antenna in [115] for all possible configurations [113].

value 1 while a part that is not connected by any edge is represented by 0. The adjacency matrix representations for all possible configurations are shown in Table 4.1. The matrices shown in Table 4.1 are intersected together and divided into sub-matrices as represented in Eq. (4.6) to Eq. (4.10). These sub-matrices represent various sections of the antenna structure that are connected by edges. For example, to achieve mode ($n = 1$), the antenna topology needs to be represented according to the first matrix of Table 4.1. This matrix is divided into six sub-matrices of the same size as shown in Eq. (4.6).

$$A_{(mode\,n=1)} = S_1 + S_2 + S_3 + S_4 + S_5 + S_6 \tag{4.6}$$

The sub-matrices composing the adjacency matrix of mode $n = 1$ are as follows:

$$S_1 = \begin{cases} a_{ij} = 1; & i = 6, 7, 8, 9, 10 \text{ and } j = 3, 4, 5, 6, 7 \\ a_{ij} = 0 & \text{otherwise} \end{cases}$$

$$= \begin{bmatrix}
0 & 0 & 0 & 0 & 0 & 0 & 0 & 0 & 0 & 0 & 0 & 0 & 0 \\
0 & 0 & 0 & 0 & 0 & 0 & 0 & 0 & 0 & 0 & 0 & 0 & 0 \\
0 & 0 & 0 & 0 & 0 & 0 & 0 & 0 & 0 & 0 & 0 & 0 & 0 \\
0 & 0 & 0 & 0 & 0 & 0 & 0 & 0 & 0 & 0 & 0 & 0 & 0 \\
0 & 0 & 0 & 0 & 0 & 0 & 0 & 0 & 0 & 0 & 0 & 0 & 0 \\
0 & 0 & 1 & 1 & 1 & 1 & 1 & 0 & 0 & 0 & 0 & 0 & 0 \\
0 & 0 & 1 & 1 & 1 & 1 & 1 & 0 & 0 & 0 & 0 & 0 & 0 \\
0 & 0 & 1 & 1 & 1 & 1 & 1 & 0 & 0 & 0 & 0 & 0 & 0 \\
0 & 0 & 1 & 1 & 1 & 1 & 1 & 0 & 0 & 0 & 0 & 0 & 0 \\
0 & 0 & 1 & 1 & 1 & 1 & 1 & 1 & 0 & 0 & 0 & 0 & 0 \\
0 & 0 & 0 & 0 & 0 & 0 & 0 & 0 & 0 & 0 & 0 & 0 & 0 \\
0 & 0 & 0 & 0 & 0 & 0 & 0 & 0 & 0 & 0 & 0 & 0 & 0 \\
0 & 0 & 0 & 0 & 0 & 0 & 0 & 0 & 0 & 0 & 0 & 0 & 0
\end{bmatrix}$$

$$S_2 = \begin{cases} a_{ij} = 1; & i = 7 \text{ and } j = 2 \\ a_{ij} = 0 & \text{otherwise} \end{cases}$$

$$= \begin{bmatrix}
0 & 0 & 0 & 0 & 0 & 0 & 0 & 0 & 0 & 0 & 0 & 0 & 0 \\
0 & 0 & 0 & 0 & 0 & 0 & 0 & 0 & 0 & 0 & 0 & 0 & 0 \\
0 & 0 & 0 & 0 & 0 & 0 & 0 & 0 & 0 & 0 & 0 & 0 & 0 \\
0 & 0 & 0 & 0 & 0 & 0 & 0 & 0 & 0 & 0 & 0 & 0 & 0 \\
0 & 0 & 0 & 0 & 0 & 0 & 0 & 0 & 0 & 0 & 0 & 0 & 0 \\
0 & 0 & 0 & 0 & 0 & 0 & 0 & 0 & 0 & 0 & 0 & 0 & 0 \\
0 & 1 & 0 & 0 & 0 & 0 & 0 & 0 & 0 & 0 & 0 & 0 & 0 \\
0 & 0 & 0 & 0 & 0 & 0 & 0 & 0 & 0 & 0 & 0 & 0 & 0 \\
0 & 0 & 0 & 0 & 0 & 0 & 0 & 0 & 0 & 0 & 0 & 0 & 0 \\
0 & 0 & 0 & 0 & 0 & 0 & 0 & 0 & 0 & 0 & 0 & 0 & 0 \\
0 & 0 & 0 & 0 & 0 & 0 & 0 & 0 & 0 & 0 & 0 & 0 & 0 \\
0 & 0 & 0 & 0 & 0 & 0 & 0 & 0 & 0 & 0 & 0 & 0 & 0 \\
0 & 0 & 0 & 0 & 0 & 0 & 0 & 0 & 0 & 0 & 0 & 0 & 0
\end{bmatrix}$$

$$S_3 = \begin{cases} a_{ij} = 1; & i = 5 \text{ and } j = 4 \\ a_{ij} = 0 & \text{otherwise} \end{cases}$$

$$= \begin{bmatrix}
0 & 0 & 0 & 0 & 0 & 0 & 0 & 0 & 0 & 0 & 0 & 0 & 0 \\
0 & 0 & 0 & 0 & 0 & 0 & 0 & 0 & 0 & 0 & 0 & 0 & 0 \\
0 & 0 & 0 & 0 & 0 & 0 & 0 & 0 & 0 & 0 & 0 & 0 & 0 \\
0 & 0 & 0 & 0 & 0 & 0 & 0 & 0 & 0 & 0 & 0 & 0 & 0 \\
0 & 0 & 0 & 1 & 0 & 0 & 0 & 0 & 0 & 0 & 0 & 0 & 0 \\
0 & 0 & 0 & 0 & 0 & 0 & 0 & 0 & 0 & 0 & 0 & 0 & 0 \\
0 & 0 & 0 & 0 & 0 & 0 & 0 & 0 & 0 & 0 & 0 & 0 & 0 \\
0 & 0 & 0 & 0 & 0 & 0 & 0 & 0 & 0 & 0 & 0 & 0 & 0 \\
0 & 0 & 0 & 0 & 0 & 0 & 0 & 0 & 0 & 0 & 0 & 0 & 0 \\
0 & 0 & 0 & 0 & 0 & 0 & 0 & 0 & 0 & 0 & 0 & 0 & 0 \\
0 & 0 & 0 & 0 & 0 & 0 & 0 & 0 & 0 & 0 & 0 & 0 & 0 \\
0 & 0 & 0 & 0 & 0 & 0 & 0 & 0 & 0 & 0 & 0 & 0 & 0 \\
0 & 0 & 0 & 0 & 0 & 0 & 0 & 0 & 0 & 0 & 0 & 0 & 0
\end{bmatrix}$$

$$S_4 = \begin{cases} a_{ij} = 1; & i = 4 \text{ and } j = 4, 5, 6, 7 \text{ or } i = 5 \text{ and } j = 5, 6, 7 \\ a_{ij} = 0 & \text{otherwise} \end{cases}$$

$$= \begin{bmatrix}
0 & 0 & 0 & 0 & 0 & 0 & 0 & 0 & 0 & 0 & 0 & 0 & 0 \\
0 & 0 & 0 & 0 & 0 & 0 & 0 & 0 & 0 & 0 & 0 & 0 & 0 \\
0 & 0 & 0 & 0 & 0 & 0 & 0 & 0 & 0 & 0 & 0 & 0 & 0 \\
0 & 0 & 0 & 1 & 1 & 1 & 1 & 0 & 0 & 0 & 0 & 0 & 0 \\
0 & 0 & 0 & 0 & 1 & 1 & 1 & 0 & 0 & 0 & 0 & 0 & 0 \\
0 & 0 & 0 & 0 & 0 & 0 & 0 & 0 & 0 & 0 & 0 & 0 & 0 \\
0 & 0 & 0 & 0 & 0 & 0 & 0 & 0 & 0 & 0 & 0 & 0 & 0 \\
0 & 0 & 0 & 0 & 0 & 0 & 0 & 0 & 0 & 0 & 0 & 0 & 0 \\
0 & 0 & 0 & 0 & 0 & 0 & 0 & 0 & 0 & 0 & 0 & 0 & 0 \\
0 & 0 & 0 & 0 & 0 & 0 & 0 & 0 & 0 & 0 & 0 & 0 & 0 \\
0 & 0 & 0 & 0 & 0 & 0 & 0 & 0 & 0 & 0 & 0 & 0 & 0 \\
0 & 0 & 0 & 0 & 0 & 0 & 0 & 0 & 0 & 0 & 0 & 0 & 0 \\
0 & 0 & 0 & 0 & 0 & 0 & 0 & 0 & 0 & 0 & 0 & 0 & 0
\end{bmatrix}$$

$$S_5 = \begin{cases} a_{ij} = 1; & i = 5 \text{ and } j = 3 \\ a_{ij} = 0 & \text{otherwise} \end{cases}$$

$$= \begin{bmatrix} 0 & 0 & 0 & 0 & 0 & 0 & 0 & 0 & 0 & 0 & 0 & 0 & 0 \\ 0 & 0 & 0 & 0 & 0 & 0 & 0 & 0 & 0 & 0 & 0 & 0 & 0 \\ 0 & 0 & 0 & 0 & 0 & 0 & 0 & 0 & 0 & 0 & 0 & 0 & 0 \\ 0 & 0 & 0 & 0 & 0 & 0 & 0 & 0 & 0 & 0 & 0 & 0 & 0 \\ 0 & 0 & 1 & 0 & 0 & 0 & 0 & 0 & 0 & 0 & 0 & 0 & 0 \\ 0 & 0 & 0 & 0 & 0 & 0 & 0 & 0 & 0 & 0 & 0 & 0 & 0 \\ 0 & 0 & 0 & 0 & 0 & 0 & 0 & 0 & 0 & 0 & 0 & 0 & 0 \\ 0 & 0 & 0 & 0 & 0 & 0 & 0 & 0 & 0 & 0 & 0 & 0 & 0 \\ 0 & 0 & 0 & 0 & 0 & 0 & 0 & 0 & 0 & 0 & 0 & 0 & 0 \\ 0 & 0 & 0 & 0 & 0 & 0 & 0 & 0 & 0 & 0 & 0 & 0 & 0 \\ 0 & 0 & 0 & 0 & 0 & 0 & 0 & 0 & 0 & 0 & 0 & 0 & 0 \\ 0 & 0 & 0 & 0 & 0 & 0 & 0 & 0 & 0 & 0 & 0 & 0 & 0 \\ 0 & 0 & 0 & 0 & 0 & 0 & 0 & 0 & 0 & 0 & 0 & 0 & 0 \end{bmatrix}$$

$$S_6 = \begin{cases} a_{ij} = 1; & i = 4 \text{ and } j = 3 \\ a_{ij} = 0 & \text{otherwise} \end{cases}$$

$$= \begin{bmatrix} 0 & 0 & 0 & 0 & 0 & 0 & 0 & 0 & 0 & 0 & 0 & 0 & 0 \\ 0 & 0 & 0 & 0 & 0 & 0 & 0 & 0 & 0 & 0 & 0 & 0 & 0 \\ 0 & 0 & 0 & 0 & 0 & 0 & 0 & 0 & 0 & 0 & 0 & 0 & 0 \\ 0 & 0 & 1 & 0 & 0 & 0 & 0 & 0 & 0 & 0 & 0 & 0 & 0 \\ 0 & 0 & 0 & 0 & 0 & 0 & 0 & 0 & 0 & 0 & 0 & 0 & 0 \\ 0 & 0 & 0 & 0 & 0 & 0 & 0 & 0 & 0 & 0 & 0 & 0 & 0 \\ 0 & 0 & 0 & 0 & 0 & 0 & 0 & 0 & 0 & 0 & 0 & 0 & 0 \\ 0 & 0 & 0 & 0 & 0 & 0 & 0 & 0 & 0 & 0 & 0 & 0 & 0 \\ 0 & 0 & 0 & 0 & 0 & 0 & 0 & 0 & 0 & 0 & 0 & 0 & 0 \\ 0 & 0 & 0 & 0 & 0 & 0 & 0 & 0 & 0 & 0 & 0 & 0 & 0 \\ 0 & 0 & 0 & 0 & 0 & 0 & 0 & 0 & 0 & 0 & 0 & 0 & 0 \\ 0 & 0 & 0 & 0 & 0 & 0 & 0 & 0 & 0 & 0 & 0 & 0 & 0 \\ 0 & 0 & 0 & 0 & 0 & 0 & 0 & 0 & 0 & 0 & 0 & 0 & 0 \end{bmatrix}$$

Thus, applying Eq. (4.6) gives us:

$$A = \begin{bmatrix} 0 & 0 & 0 & 0 & 0 & 0 & 0 & 0 & 0 & 0 & 0 & 0 & 0 \\ 0 & 0 & 0 & 0 & 0 & 0 & 0 & 0 & 0 & 0 & 0 & 0 & 0 \\ 0 & 0 & 0 & 0 & 0 & 0 & 0 & 0 & 0 & 0 & 0 & 0 & 0 \\ 0 & 0 & 1 & 1 & 1 & 1 & 1 & 0 & 0 & 0 & 0 & 0 & 0 \\ 0 & 0 & 1 & 1 & 1 & 1 & 1 & 0 & 0 & 0 & 0 & 0 & 0 \\ 0 & 0 & 1 & 1 & 1 & 1 & 1 & 0 & 0 & 0 & 0 & 0 & 0 \\ 0 & 1 & 1 & 1 & 1 & 1 & 1 & 0 & 0 & 0 & 0 & 0 & 0 \\ 0 & 0 & 1 & 1 & 1 & 1 & 1 & 0 & 0 & 0 & 0 & 0 & 0 \\ 0 & 0 & 1 & 1 & 1 & 1 & 1 & 0 & 0 & 0 & 0 & 0 & 0 \\ 0 & 0 & 1 & 1 & 1 & 1 & 1 & 0 & 0 & 0 & 0 & 0 & 0 \\ 0 & 0 & 0 & 0 & 0 & 0 & 0 & 0 & 0 & 0 & 0 & 0 & 0 \\ 0 & 0 & 0 & 0 & 0 & 0 & 0 & 0 & 0 & 0 & 0 & 0 & 0 \\ 0 & 0 & 0 & 0 & 0 & 0 & 0 & 0 & 0 & 0 & 0 & 0 & 0 \end{bmatrix} = \sum_{i=1}^{6} S_i$$

Table 4.1: Adjacency matrix representation for all possible antenna configurations [113]

Mode n=1	Mode n=1 With 90°	Mode n=2	Mode n=2 with 45°	Mode n=0

The adjacency matrices representing the antenna operation in other modes are represented below in the Eqs. (4.7), (4.8), (4.9), and (4.10). The same procedure is used for all modes of operations. The sub-matrices S_{ij} for each of the adjacency matrices are defined in Table 4.2.

$$A_{(\text{mode } n=1 \text{ Rotated } 90°)} = S_1 + S_2 + S_3 + S_8 + S_{12} + S_{13} + S_{14} \tag{4.7}$$

$$\begin{aligned} A_{(\text{mode } n=2)} = {} & S_1 + S_2 + S_3 + S_4 + S_5 + S_6 + S_7 + S_8 + S_9 + S_{10} \\ & + S_{15} + S_{17} + S_{27} \end{aligned} \tag{4.8}$$

$$\begin{aligned} A_{(\text{mode } n=2 \text{ Rotated } 45°)} = {} & S_1 + S_2 + S_3 + S_4 + S_5 + S_8 + S_9 + S_{13} + S_{14} \\ & + S_{15} + S_{17} + S_{19} + S_{20} + S_{21} + S_{22} + S_{23} + S_{24} \end{aligned} \tag{4.9}$$

$$A_{(\text{mode } n=0)} = \sum_{i=1}^{27} S_i \tag{4.10}$$

Each sub-matrix in Table 4.2 that constitutes the result of intersection between various adjacency matrices is translated into a graph model that represents an antenna section. Thus, we have 27 different antenna sections as shown in Fig. 4.17. Inside each section, the square patches are

Table 4.2: The sub-matrices composing the matrices of Table 4.1 [113]

$S1 = \begin{cases} aij = 1; \ i = 6,7,8,9,10 \ and \ j = 3,4,5,6,7 \\ aij = 0 \ Otherwise \end{cases}$	$S2 = \begin{cases} aij = 1; \ i = 7 \ and \ j = 2 \\ aij = 0 \ Otherwise \end{cases}$	$S3 = \begin{cases} aij = 1; \ i = 5 \ and \ j = 4 \\ aij = 0 \ Otherwise \end{cases}$
$S4 = \begin{cases} aij = 1; \ i = 4 \ and \ j = 4,5,6,7 \\ or \ i = 5 \ and \ j = 5,6,7 \\ aij = 0 \ Otherwise \end{cases}$	$S5 = \begin{cases} aij = 1; \ i = 5 \ and \ j = 3 \\ aij = 0 \ Otherwise \end{cases}$	$S6 = \begin{cases} aij = 1; \ i = 4 \ and \ j = 3 \\ aij = 0 \ Otherwise \end{cases}$
$S7 = \begin{cases} aij = 1; \ i = 2 \ and \ j = 5,6,7 \\ or \ i = 3 \ and \ j = 3,4,5,6,7,8,9 \\ or \ i = 3 \ and \ j = 3 \\ aij = 0 \ Otherwise \end{cases}$	$S8 = \begin{cases} aij = 1; \ i = 6 \ and \ j = 2 \ or \ i = 7 \ and \ j = \\ or \ i = 8 \ and \ j = 1,2 \ or \ i = 9 \ and \ j = 2 \\ or \ i = 10 \ and \ j = 2 \\ aij = 0 \ Otherwise \end{cases}$	$S9 = \begin{cases} aij = 1; \ i = 5 \ and \ j = 2 \\ aij = 0 \ Otherwise \end{cases}$
$S10 = \begin{cases} aij = 1; \ i = 4 \ j = 3 \\ aij = 0 \ Otherwise \end{cases}$	$S11 = \begin{cases} aij = 1; \ i = 5 \ j = 1 \\ aij = 0 \ Otherwise \end{cases}$	$S12 = \begin{cases} aij = 1; \ i = 6 \ j = 1 \\ aij = 0 \ Otherwise \end{cases}$
$S13 = \begin{cases} aij = 1; \ i = 9 \ j = 1 \\ aij = 0 \ Otherwise \end{cases}$	$S14 = \begin{cases} aij = 1; \ i = 10 \ j = 1 \\ aij = 0 \ Otherwise \end{cases}$	$S15 = \begin{cases} aij = 1; \ j = 8,9 \ and \ i = 5,6,7,8,9,10 \\ j = 10 \ and \ i = 6,7,8,9,10 \\ aij = 0 \ Otherwise \end{cases}$
$S16 = \begin{cases} aij = 1; \ i = 4 \ j = 8,9,10 \\ i = 5 \ j = 10 \\ aij = 0 \ Otherwise \end{cases}$	$S17 = \begin{cases} aij = 1; \ i = 11 \ j = 3,4,5,6 \\ i = 12 \ j = 5,6,7 \\ aij = 0 \ Otherwise \end{cases}$	$S18 = \begin{cases} aij = 1; \ i = 12 \ j = 4 \\ aij = 0 \ Otherwise \end{cases}$
$S19 = \begin{cases} aij = 1; \ i = 13 \ j = 8,7,6,5 \\ aij = 0 \ Otherwise \end{cases}$	$S20 = \begin{cases} aij = 1; \ i = 13 \ j = 4 \\ aij = 0 \ Otherwise \end{cases}$	$S21 = \begin{cases} aij = 1; \ i = 11 \ and \ j = 2 \\ i = 12 \ and \ j = 2,3 \\ aij = 0 \ Otherwise \end{cases}$
$S22 = \begin{cases} aij = 1; \ i = 11 \ j = 10 \\ aij = 0 \ Otherwise \end{cases}$	$S23 = \begin{cases} aij = 1; \ i = 7,8 \ j = 10 \\ aij = 0 \ Otherwise \end{cases}$	$S24 = \begin{cases} aij = 1; \ i = 12 \ j = 8,9 \\ aij = 0 \ Otherwise \end{cases}$
$S25 = \begin{cases} aij = 1; \ i = 1 \ j = 5,6,7,8,9 \\ or \ i = 2 \ and \ j = 8,9,10 \\ or \ i = 3 \ and \ j = 10,11 \\ or \ i = 4 \ and \ j = 11,12 \\ or \ i = 5,6,7,8,9 \ and \ j = 11,12,13 \\ or \ i = 11 \ and \ j = 11 \\ or \ i = 6 \ and \ j = 10 \ or \ i = 10 \ and \ j = 11,12 \\ or \ i = 12 \ and \ j = 10 \\ aij = 0 \ Otherwise \end{cases}$	$S26 = \begin{cases} aij = 1; \ i = 2 \ j = 4 \\ aij = 0 \ Otherwise \end{cases}$	$S27 = \begin{cases} aij = 1; \ i = 6 \ j = 2 \\ aij = 0 \ Otherwise \end{cases}$

connected constantly, which eliminates the need for switches. Switches will be used only to connect the different sections to each other. Parts belonging to the same sections are always connected, and there is no need for switches inside each section. Some antenna parts are never connected, to achieve polarization diversity, and they are shown in black in Fig. 4.17 [113].

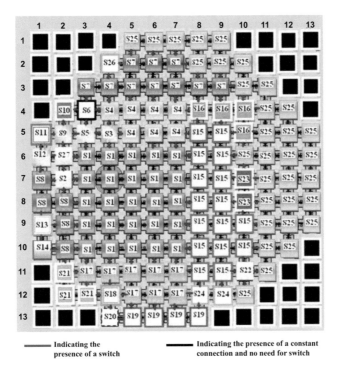

Figure 4.17: Different antenna sections in different colors. The black parts represent the idle parts (parts never connected to achieve polarization diversity).

By applying this technique, the number of switches is reduced by more than 100 switches from 312 to 166, while preserving the antenna topology. The reduction of the number of switches has decreased the biasing cost for the various switches as well as limited the power consumption of such a large antenna structure. It has also reduced the effect of these switches on the antenna operation. Preserving the antenna topology while reducing the number of switches has preserved the antenna's radiation characteristics. A comparison of the reflection coefficient between the original [115] antenna and the antenna with reduced number of switches is shown in Fig. 4.18 for the mode $n = 0$ proving that the technique does not disturb the antenna radiation characteristics. The radiation pattern at 5.83 GHz for $\Phi = 0°$ and $\Phi = 90°$ is shown in Fig. 4.19.

Example 4.4

Let's take the single-part antenna shown in Fig. 2.9 [80], and discussed in Example 2.3. The graph modeling of this antenna follows rule 2 and is shown in Fig. 2.10. The number of available

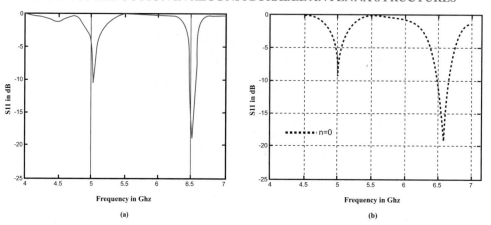

Figure 4.18: A comparison of the S11 parameter between the original antenna and the antenna with reduced number of switches for the mode $n = 0$. (a) Reduced number of switches, (b) Original antenna.

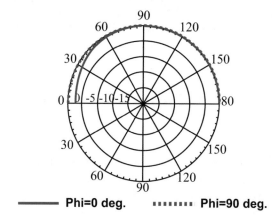

Figure 4.19: The antenna radiation pattern at 5.83 GHz for $\phi = 0$ deg, and 90 deg.

antenna configurations is thus calculated according to Eq. (4.4a) and it is:

$$NAC = \frac{N}{2} + 1 = \frac{20}{2} + 1 = 11$$

The number of configurations required in [80] is only five, so NAC should be equal to 5, so by applying Eq. (4.4b) we end up with eight vertices as shown below:

$$NAC = 5$$
$$N = 2 \times (NAC - 1) = 2 \times (5 - 1) = 8$$

These eight vertices represent the eight end points of four switches according to rule 2. The optimal design includes four switches. All of these four switches are activated at the same time and not individually. The graph model of the optimal antenna is shown in Fig. 4.20. Each vertex in this design represents an end-point of a switch in the antenna structure.

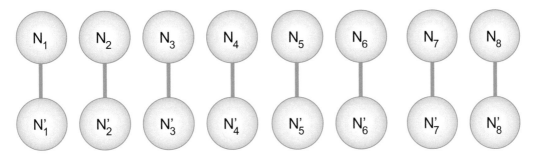

Figure 4.20: Graph of the optimized antenna topology for all possible connections.

4.4 DISCUSSION

In this chapter we present a redundancy reduction approach that is derived from the graph models of reconfigurable antennas. This redundancy reduction technique is based on the iterative design method presented in Chapter 3; however in this chapter it is formulated to provide an easy tool for antenna designers. The approach discussed herein is based on the concept that each unique path in the graph is responsible for a distinctive function which is indicative of tight constraints. By removing redundant components from the antenna structure the designer is reducing the cost of design as well as reducing numerous losses in the anticipated system. This technique can also verify previously designed antennas as shown in some of the examples of this chapter. In that case, the designer is faced with the question of whether or not to redesign the antenna's topology or preserve it while minimizing the number of unnecessary elements.

The objective of this method is to reduce the time needed for simulations of antenna structures during the design process by mathematically determining the necessary and required number of elements in an antenna structure. In reality the graph modeling of reconfigurable antennas is transforming them from bulky mechanical structures into mathematical models that are easily accessible and can be software controlled by various means.

CHAPTER 5

Analyzing the Complexity and Reliability of Switch-Frequency Reconfigurable Antennas Using Graph Models

5.1 INTRODUCTION

The incorporation of switches into reconfigurable antenna structures increases their complexity which in turn diminishes the reliability of the antennas. In particular, the reliability of reconfigurable antennas is of the upmost importance in unknown and unpredictable environments. The design of switching elements is highly dependent on environmental conditions. For instance, if reconfigurable antennas are deployed in space, the environment is unpredictable and the antenna structure is difficult to access for any maintenance activity [117].

Various publications discuss the use of switching components such as RF MEMS to reconfigure antennas [118–120]. However, there are some issues in integrating commercially packaged RF MEMS into a reconfigurable antenna as discussed in [6, 121]; not only the insertion loss and isolation behavior of the switches are addressed, but also their impact on the radiation characteristics of the antenna.

Environmental effects, such as carbon contamination [122], have an effect on the reliability of RF MEMS. Most publications in this area do not reflect the reliability of switch-reconfigurable antennas, and few designers investigate the environmental effects on good system operation.

The fundamentals of improving systems reliability is first addressed by Shannon and Moore. They propose using redundancy to increase reliability [123]. The reliability is also discussed in [124], where the fundamental mathematics of fault-tolerant circuit-switching networks is illustrated. Both [123, 124] emphasize and recommend redundancy to improve the reliability of any switching circuit. This work is based on electronic circuits without considering any electromagnetic aspect. In this chapter we base our discussion on finding a trade-off between reduction in redundancy and maintaining reliability. The analysis done in [123] is used to study the effect of the complexity reduction approach presented previously in Chapter 4 on the reliability of a particular antenna.

5.2   GRAPH MODELING EQUIVALENT ANTENNA CONFIGURATIONS

Graph modeling is a useful tool for analyzing reconfigurable antennas as discussed in this book as well as in [47, 113, 125–127]. There are several ways to graph model reconfigurable antennas as discussed in Chapter 2. In a switch frequency reconfigurable antenna, many switching configurations yield the same antenna frequency behavior without affecting the other radiation properties. These configurations are called equivalent configurations. Equivalent configurations are obtained from simulations. These equivalent configurations constitute back-up configurations for maintaining the same antenna performance at a certain frequency. Graph models are used to represent the equivalent configurations for an antenna structure at a given frequency.

For example, the antenna shown in Fig. 4.9 [101] resonates at 5 GHz for eight different switch configurations as shown in Fig. 5.1. These eight configurations are summarized in Table 5.1. These eight configurations achieve the desired resonance frequency at 5 GHz in addition to other resonant frequencies as shown in Fig. 5.2. It is true that reducing the number of switches decreases the number of equivalent antenna configurations at each resonant frequency [47, 113]. However, the number of remaining configurations is sufficient for reliable antenna operation [117].

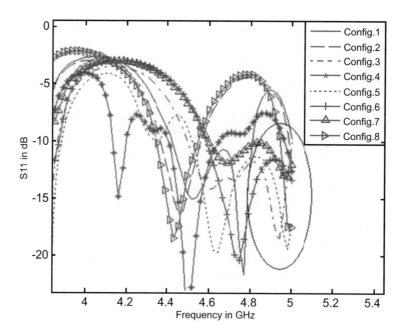

Figure 5.1: S11 plot for the antenna in [101] for the configurations presented in Table 5.1, zoomed in at 5 GHz [117].

Table 5.1: The different configurations of the antenna in [101] that lead to operation at 5 GHz [117]

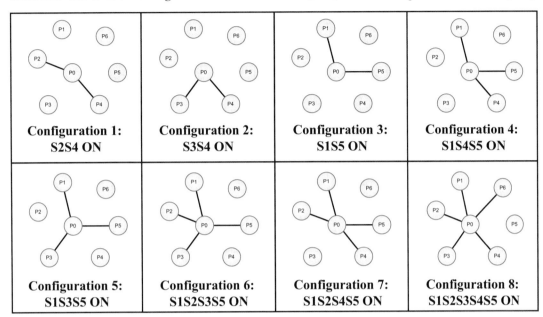

Another example can be the previously optimized antenna (Chapter 4) shown in Fig. 5.3. Some of the antenna configurations for different antenna resonances are shown in Table 5.2.

Table 5.2: Some antenna configurations for different resonances (all frequencies are in GHz)

Even after optimization, this antenna has several equivalent configurations for each resonant frequency. The optimization technique reduced the number of switches and hence cost without compromising the performance of the system at operating frequencies [117]. Even after optimization, we still have enough equivalent configurations for a good continuous antenna operation.

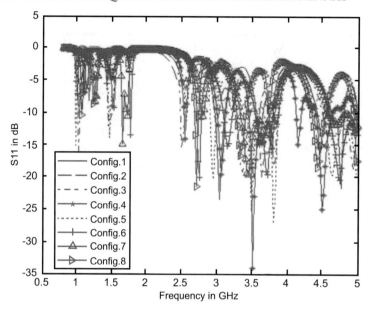

Figure 5.2: S11 plot for the antenna in [101] showing multi-band operation [117].

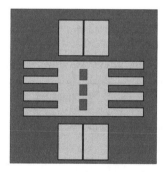

Figure 5.3: The Optimized Antenna from Chapters 3 and 4 [117].

It is also important to point out that certain resonant frequencies are only achievable with a single configuration. In this case, there is a need to develop some methods to improve the efficiency and insure continuous antenna operation.

In the next section the complexity and reliability of reconfigurable antennas are formulated and methods for improving the antenna reliability are proposed [117].

5.3   RELIABILITY FORMULATION FOR FREQUENCY RECONFIGURABLE ANTENNAS

According to Shannon and Moore's terminology [123], a switch failure occurs when:

1. A switch is originally OFF and fails to switch ON upon request

2. A switch is originally ON and fails to switch OFF upon request

A switching failure heavily affects the reliability of a switch reconfigurable antenna. These failures are due to the environment of operation, the aging and corrosion process, and the frequency of operation. Thus, the reliability of reconfigurable antennas depends on all the previously mentioned factors.

With graphs, we can calculate the reliability using models which represent the different antenna configurations. The reliability is dependent on the number of antenna configurations at a certain frequency and the probability to achieve these configurations. However, it is inversely proportional to the number of edges needed to create these configurations. The solution is to design reconfigurable antennas with several equivalent configurations but only a small number of connections (edges). The reliability can be expressed as:

$$R(f) = \frac{\sum\limits_{i=1}^{Nc(f)} \sum\limits_{j=1}^{NE_i(f)} P(E_{ij})}{\sum\limits_{i=1}^{Nc(f)} NE_i(f)} \times 100 \tag{5.1}$$

where:

$R(f)$ = The reconfigurable antenna reliability at a particular frequency f
$N_C(f)$ = The number of equivalent configurations achieving the frequency f
$NE(f)$ = The number of edges for different equivalent configurations at the frequency f
$P(E)$ = Probability of achieving the edge $E = 1-$Probability of a switch failing

Example 5.1

Let us consider the antenna shown in Fig. 5.3. Assume we want to calculate the antenna's reliability at 2.9 GHz. According to Table 5.2, at 2.9 GHz the antenna has three equivalent configurations which resonate at this particular frequency. Assuming the probability that each edge exists in a given configuration is equal to 0.98 (picked here as an example), this is the probability of success, then according to Eq. (5.1):

$$R(2.9 \text{ GHz}) = \frac{\displaystyle\sum_{i=1}^{3} \sum_{j=1}^{NE_i(2.9)} P(E_{ij})}{\displaystyle\sum_{i=1}^{3} NE_i(2.9)} \times 100$$

$$= \frac{\displaystyle\sum_{j=1}^{NE_1(2.9)} P(E_{1j}) + \sum_{j=1}^{NE_2(2.9)} P(E_{2j}) + \sum_{j=1}^{NE_3(2.9)} P(E_{3j})}{NE_1(2.9) + NE_2(2.9) + NE_3(2.9)} \times 100$$

$$= \frac{P(E_{11}) + P(E_{12}) + P(E_{21}) + P(E_{22}) + P(E_{23}) + P(E_{31}) + P(E_{32}) + P(E_{33}) + P(E_{34})}{2 + 3 + 4}$$
$$\times 100$$

$$= \frac{9 \times 0.98}{9} \times 100 = 98\%$$

Example 5.2

Let us now consider the same antenna shown in Fig. 5.3 but at 1.7 GHz. Let's assume that the probability of switching success with switch 1 is 0.999, the probability of success with switch 2 is 0.998, and the probability of success with switch 3 is 0.900. According to Eq. (5.1), the reliability at 1.7 GHz is:

$$R(1.7 \text{ GHz}) = \frac{\displaystyle\sum_{i=1}^{1} \sum_{j=1}^{NE_i(1.7)} P(E_{ij})}{\displaystyle\sum_{i=1}^{3} NE_i(1.7)} \times 100$$

$$= \frac{\displaystyle\sum_{j=1}^{NE_1(1.7)} P(E_{1j})}{NE_1(1.7)} \times 100$$

$$= \frac{P(E_{11}) + P(E_{12}) + P(E_{13})}{3} \times 100$$

$$= \frac{0.999 + 0.998 + 0.900}{3} \times 100 = 96\%$$

The variation of the reliability for different probability values at a particular frequency is linear and Fig. 5.4 shows this variation at $f = 1.7$ GHz [117]. Here, if no switches are used for achieving a certain reconfiguration, then the reliability is 100%. An example of such a 100% reliable antenna is the antenna in Fig. 5.3 at 5.2 GHz. One of the configurations which resonate at this frequency does not use any switch.

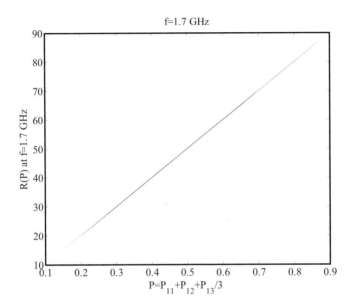

Figure 5.4: The variation of the reliability for different values of the probabilities at $f = 1.7$ GHz [117].

5.4   GENERAL COMPLEXITY OF RECONFIGURABLE ANTENNAS

Increasing the number of edges in a reconfigurable antenna graph model adds to the complexity of the antenna. The complexity is based on the size of the graph, i.e., the number of edges for all possible connections in that graph [117].

$$C = NE \qquad (5.2)$$

where NE represents the number of edges for all possible connections in a given configuration.

Removal of redundant elements results in reduction of the general complexity of the hardware used as well as simplification of software analysis employed to control the reconfiguration technique.

Example 5.3

For example, the antenna discussed in Example 4.3, is optimized while preserving its core function and its topology. It is concluded that some of the switches are not needed to achieve the required functions. This reduction in switches reduces the complexity of the antenna. The general complexity of this antenna before optimization is $C = NE = 312$, and after optimization C is reduced to $C = 166$.

5.5   CORRELATION BETWEEN COMPLEXITY AND RELIABILITY OF RECONFIGURABLE ANTENNAS

An antenna can have several configurations at different frequencies of operation, therefore it is essential to define the complexity at each particular frequency. Eq. (5.3) defines this frequency-dependent complexity.

$$C(f) = \max_{i=1,NC(f)} (NE_i(f)) \tag{5.3}$$

where:

$\quad\quad C(f)\quad$ represents the complexity of the antenna system at a frequency f
$\quad NC(f)\quad$ represents the number of equivalent configurations at a frequency f
$\quad NE_i(f)\quad$ represents the number of edges at the configuration i for a frequency f

As an example, let's take the antenna shown in Fig. 5.1. The different configurations of this antenna at 5 GHz are shown in Table 5.1. The complexity of this antenna at 5 GHz is calculated by Eq. (5.3) as:

$$C(5 \text{ GHz}) = \max_{i=1,8}(NE_i(5)) = \text{Max}(2, 2, 2, 3, 3, 4, 4, 5) = 5$$

The complexity of the optimized antenna shown in Fig. 5.3 at 5.2 GHz is calculated by Eq. (5.3) as:

$$C(5.2 \text{ GHz}) = \max_{i=1,7}(NE_i(5.2)) = \text{Max}(1, 1, 1, 2, 2, 2, 0) = 2$$

The correlation between the complexity of an antenna at a frequency f and its reliability at that same frequency can be derived using Eq. (5.2). The denominator of the reliability equation (Eq. (5.1)) contains the total number of edges for all configurations at a particular frequency $\left(\sum_{i=1}^{N_C(f)} NE_i(f)\right)$. This total number of edges includes configuration with the maximum number of edges. This configuration reveals the frequency dependent complexity as expressed in Eq. (5.3). Therefore, the denominator of Eq. (5.1) can be expressed as:

$$\sum_{i=1}^{N_C(f)} NE_i(f) = C(f) + \sum_{k=1}^{N'_C(f)} NE_K(f) \tag{5.4}$$

where $C(f)$ is calculated in Eq. (5.3) and $N'C(f)$ is the number of equivalent configurations at a frequency f without the configuration with maximum edges.

The reliability can then be expressed as in Eq. (5.5)

$$R(f) = \frac{\displaystyle\sum_{i=1}^{Nc(f)} \sum_{j=1}^{NE_i(f)} P(E_{ij})}{C(f) + \displaystyle\sum_{k=1}^{N'c(f)} NE_k(f)} \times 100 \tag{5.5}$$

From Eq. (5.5) we can deduce that the reliability of a reconfigurable antenna at a frequency f is inversely proportional to the complexity of that antenna's structure at the same frequency f.

Example 5.4

Taking the same antenna from Example 5.1 and recalculating the reliability according to Eq. (5.5), reveals:

$$C(2.9 \text{ GHz}) = \underset{i=1,3}{\text{Max}}(NE_i(2.9)) = \text{Max}(2, 3, 4) = 4$$

$$R(2.9 \text{ GHz}) = \frac{\displaystyle\sum_{i=1}^{3} \sum_{j=1}^{NE_i(2.9)} P(E_{ij})}{C(2.9) + \displaystyle\sum_{k=1}^{2} NE_k(2.9)} \times 100$$

$$= \frac{\displaystyle\sum_{j=1}^{NE_1(2.9)} P(E_{1j}) + \sum_{j=1}^{NE_2(2.9)} P(E_{2j}) + \sum_{j=1}^{NE_3(2.9)} P(E_{3j})}{4 + NE_1(2.9) + NE_2(2.9)} \times 100$$

$$= \frac{P(E_{11}) + P(E_{12}) + P(E_{21}) + P(E_{22}) + P(E_{23}) + P(E_{31}) + P(E_{32}) + P(E_{33}) + P(E_{34})}{4 + 2 + 3}$$

$$\times 100$$

$$= \frac{9 \times 0.98}{9} \times 100 = 98\%$$

5.6   INCREASING THE RELIABILITY OF RECONFIGURABLE ANTENNAS

In this section we propose two methods to improve the reliability of a reconfigurable antenna system. These methods are based on the presence of antenna-redundant configurations even after the implementation of the redundancy reduction approach. These redundant configurations are a manifestation of the antenna electromagnetic behavior under the remaining switch states. These methods utilize redundant configurations to improve the reliability of the reconfigurable antenna.

The first method suggests that one should organize the desired frequencies starting from high priority to low priority. If the first method is not sufficient to increase the reliability of an antenna at a certain desired frequency then the second method is applied. This method, based on adding a back-up switch, utilizes the analysis in [123] to improve the reliability.

Method 1: The no-switch configuration

The first method advises the designer to first prioritize the frequencies needed. The frequency with the highest priority should have more than one equivalent configuration. If we look at Table 5.2, we can deduce that the frequency with the largest number of equivalent configurations is $f = 5.2$ GHz. It has seven equivalent configurations, including the no-switch configuration (all switches off). A good design approach is to design the antenna to operate at the most important frequency or frequencies with all switches off. In that case, under the worst possible scenario of all switches breaking down at the same time, the most important frequency is always achievable.

Method 2: The back-up switch

This method proposes installing a back-up switch. The back-up switch can be installed at any place in the antenna system as long as its presence achieves the desired frequency. This method is used when a certain frequency is needed at all times, and the design does not include enough back-up configurations. Many factors come into play when installing a back-up switch. We assume that the probability of failure of a switch remains constant in time and does not change. Thus, the back-up switch method can be used if and only if it satisfies the following constraints:

1. Its probability of failure is lower than or equal to the lowest probability of failure among all switches.

2. The sum of probabilities of success in the back-up configuration is higher than the sum of probabilities of success in the original configuration.

Example 5.5

The optimized antenna in Fig. 5.3 operates at 2.05 GHz for only one configuration (S1 ON) as shown in Table 5.2. Installing a back-up switch as shown in Fig. 5.5 and activating switches S2 and S3 constitutes a back-up configuration. The graph model of this system is represented in Fig. 5.6 where P0 is replaced by P'0 since by adding the back-up switch new vertices appear and the topology of the antenna section represented by P0 has changed. As stated, the placement of the switch is up to the designer as long as the presence of that switch achieves the desired function.

The reflection coefficient plot is shown in Fig. 5.7. The reliability of this antenna can be increased by applying either or both of the two methods proposed in this section.

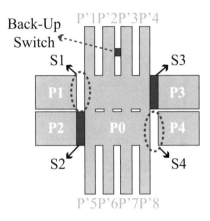

Figure 5.5: Antenna using backup switch for 2.05 GHz equivalent configurations [117].

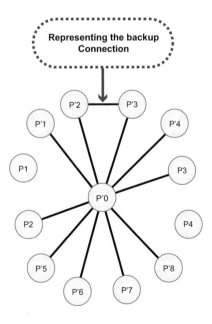

Figure 5.6: Graph model of the equivalent configuration in Fig. 5.5 [117].

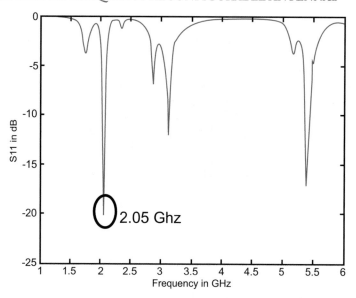

Figure 5.7: The antenna's reflection coefficient with backup switch S2 and S3 activated showing clear operation at 2.05 GHz [117].

This antenna originally has only one configuration that achieves 2.05 GHz; that configuration requires only the activation of S1. Assuming the probability of switch activation success is 0.98 for switch 1, 0.985 for switch 2, and 0.999 for switch 3, then the reliability of the antenna at $f = 2.05$ GHz is:

$$R(2.05 \text{ GHz}) = \frac{\sum_{i=1}^{1} \sum_{j=1}^{NE_i(2.05)} P(E_{ij})}{\sum_{i=1}^{3} NE_i(2.05)} \times 100 = \frac{P(E_{11})}{1} \times 100$$

$$= \frac{0.98}{1} \times 100 = 98\%$$

The back-up switch configuration adds a new configuration possibility to achieve 2.05 GHz by switching ON S2, S3 and the back-up switch. Calculating the reliability with the back-up switch

ON gives us:

$$R(2.05 \text{ GHz}) = \frac{\sum\limits_{i=1}^{2} \sum\limits_{j=1}^{NE_i(2.05)} P(E_{ij})}{\sum\limits_{i=1}^{3} NE_i(2.05)} \times 100 = \frac{\sum\limits_{j=1}^{NE_1(2.05)} P(E_{1j}) + \sum\limits_{j=1}^{NE_2(2.05)} P(E_{2j})}{NE_1(2.05) + NE_2(2.05)} \times 100$$

$$= \frac{P(E_{11}) + P(E_{21}) + P(E_{22}) + P(E_{23})}{1 + 3} \times 100$$

$$= \frac{0.98 + 0.985 + 0.999 + 0.999}{4} \times 100$$

$$= 99.075\%$$

Thus, adding the back-up switch, not only assures the continuous functionality of this antenna, but also improves its reliability at that particular frequency [117]. In this way decreasing the antenna's complexity did not affect its reliability if the methods for reliability preservation are applied. The reliability cannot only be preserved but also increased with efficient and optimal antenna designs. In the next section we present an algorithm that insures the reliability of an antenna system under unknown conditions.

5.7   RELIABILITY ASSURANCE ALGORITHM

To overcome a switch failure and thus restore a lost resonance, the following methodology is proposed. This methodology identifies the defected switch, specifies the desired frequency, and changes the antenna topology to restore the desired resonance based on the equivalent configuration.

Before applying this approach, the designer creates a library similar to Table 5.2 in which all equivalent configurations for all desired frequencies are identified. Software and learning algorithms can be used to speed the library assembly process for large antenna structures. The designer should include in the library the backup switch configuration if such configuration exists for specific frequencies. In the library table the rows correspond to the various frequencies and the columns correspond to the equivalent configurations. The algorithm is described below and shown in Fig. 5.8

- Step 1: Identify defected switch. In this step the designer identifies which switch has failed exactly.

- Step 2: Identify the desired lost frequency, where the desired frequency is the resonance at which the antenna operation is required. At this point the frequency that needs to be restored is identified.

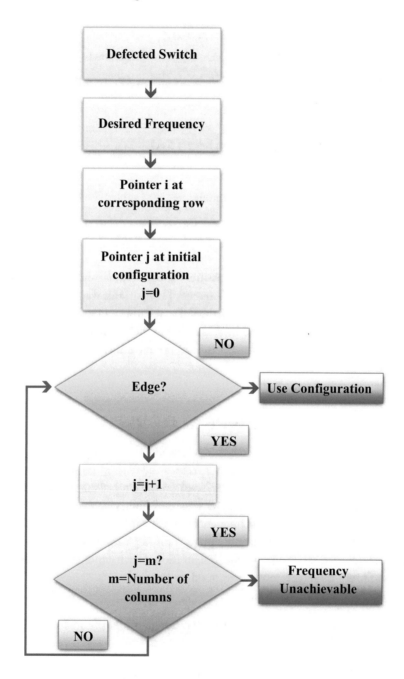

Figure 5.8: A schematic representation of the reliability algorithm.

- Step 3: In the table created, create a pointer at row i corresponding to the desired frequency. This pointer indicates the frequency that needs restoration.

- Step 4: In the table, create a pointer j at the first configuration. ($j = 0$). Pointer j indicates the various columns corresponding to the equivalent configurations at the desired frequency. The pointer j is initialized at first to point to the first configuration existing at the row of the desired frequency.

- Step 5: At j's position (first configuration) search for a possible edge representing a connection from the defected switch.

- Step 6: If no connection is found, use configuration in the column j.

- Step 7: If a connection is found move j to the next column ($j = j + 1$).

- Step 8: Are there any more configurations? Has j reached the last column?

- Step 9: If NO, repeat steps 5, 6, 7, and 8.

- Step 10: If Yes, and no solution is found, declare frequency unachievable.

5.8 DISCUSSION

In this chapter we have addressed the complexity and the reliability of reconfigurable antennas and their inter-correlation. Concerns regarding the effect of redundancy reduction on the reliability of reconfigurable antennas are addressed. It is proven that a higher complexity affects the reliability of reconfigurable antennas negatively. Solutions to preserve and increase the reliability of reconfigurable antennas in unknown environments are discussed and presented. The use of graph models allows us not to only to reduce the antenna's complexity but to accurately estimate its reliability based on corresponding edges probabilities.

CHAPTER 6

Complexity Versus Reliability in Arrays of Reconfigurable Antennas

6.1 INTRODUCTION

In the previous chapters of this book we have discussed the various implementations of graph modeling a reconfigurable antenna structure. It is proven that a graph is used as an abstract model to represent physical structures. Thus, graph modeling reconfigurable antennas transform them into software accessible devices that are easy to optimize, control, and automate. In Chapters 3 and 4 we have shown that the modeling of reconfigurable antennas using graphs leads to a redundancy reduction approach that eliminates unnecessary components from the antenna structure. Thus, graph models are utilized to formulate a reconfigurable antenna's complexity as discussed in Chapter 5. The overall complexity of an antenna system increases with the number of switches such as p-i-n diodes, RF MEMS, varactors, or optical switches employed. The reliability of the reconfigurable antenna in question is also addressed in Chapter 5 and its correlation with the complexity is shown and detailed.

In this chapter we expand previous formulations done on single element reconfigurable antennas to evaluate the complexity and reliability of reconfigurable antenna arrays using graphs. This is essential to address the continuous functioning of these arrays in unknown conditions and environments, especially that reconfigurable antenna arrays are considered as suitable candidates for implementation on various terrestrial and space communication applications. In fact, there is a huge benefit in designing reconfigurable antenna arrays for various wireless communication applications. In [128], the impact of reconfigurable antenna arrays on cognitive radio is addressed, while in [56] a MIMO antenna array system is proposed. Most reconfigurable antenna array designers resort to p-i-n diodes [129] or RF MEMS [130] as switching elements to achieve reconfiguration. Failures in such arrays are mostly due to switch failures.

Failures due to various causes, whether in the open/close behavior of the switches and its effect on the antenna radiation pattern, or antenna tuning malfunction as a result of carbon contamination and its effect on the incorporated switches, are all discussed in literature such as [6, 122, 131].

The graph modeling of reconfigurable antenna arrays is first presented. The method of graph modeling a reconfigurable antenna array is different than a single-element reconfigurable antenna.

The use of graph models to formulate arrays' complexity, the impact of the complexity on the reliability of switch reconfigurable antenna arrays, and their configuration complexity, are discussed and formulated. A technique is also proposed to improve the performance of reconfigurable antenna arrays. This technique is based on rearranging antenna configurations to ensure higher reliability. The improvement that such rearrangement introduces to the design efficiency and operation is discussed. The practicality and design aspects are shown to give the antenna researchers a concrete tool in their designing process.

6.2   THE GRAPH MODELING OF ARRAYS OF RECONFIGURABLE ANTENNAS

In an antenna array, all elements are fed through a feeding network and are placed at certain spacing from each other [56, 130]. As an example, let us consider the antenna array shown in Fig. 6.1. The array is composed of three layers. The bottom layer constitutes a common ground plane for the different elements. The middle layer constitutes the substrate Taconic TLX with a dielectric constant $\epsilon_r = 2.55$ and height 2.9 mm. The top layer constitutes the different element patches as well as the corporate feeding network. Each element is a rectangular patch with two rectangular slots dividing it into two sections connected constantly. Two switches are placed in each element to bridge over the upper and lower parts of the slots. The end-points of each switch are indicated by the nodes $\{P_{ij}, P'_{ij}\}$ (Eq. 6.1) where i refers to the element number and j refers to the switch position. For example, $P_{21} P'_{21}$ represent the end-points of the lower switch (S4) in element 2. The array's reflection coefficient for two different switch combinations is shown in Fig. 6.2. The array's fabricated prototype is shown in Fig. 6.3 with its corresponding dimensions. A comparison between the measured and simulated reflection coefficient of the array antenna when S1 is ON is shown in Fig. 6.4 indicating good analogy [132].

$$\{P_{ij}, P'_{ij}\} : \begin{cases} i = 1, j = 1 & \text{Element 1, upper switch, Switch 1 (S1)} \\ i = 1, j = 2 & \text{Element 1, lower switch, Switch 2 (S2)} \\ i = 2, j = 1 & \text{Element 2, upper switch, Switch 3 (S3)} \\ i = 2, j = 2 & \text{Element 2, lower switch, Switch 4 (S4)} \end{cases} \tag{6.1}$$

The graph model of this array is shown in Fig. 6.5 where each vertex represents the two end-points of the switches in addition to the intersection of any two lines forming the array structure. This graph model is a general representation of an antenna array where all the parameters that go into the antenna array design are taken into consideration. The general graph modeling of a switch reconfigurable antenna array is based on the fact that in an antenna array design every component in a unit cell propagates over to the other elements. The design of these arrays takes into consideration the various unit cell dimensions as well as the spacing between the various elements, in order to control and steer the array's radiation properties, as well as to control the coupling and optimize its constructive effect.

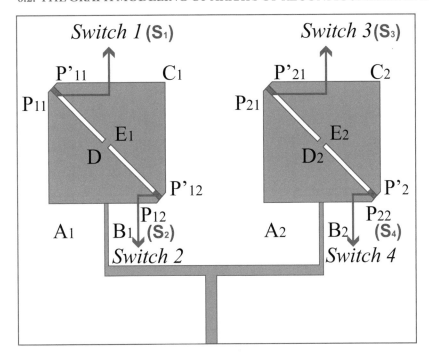

Figure 6.1: A two-element reconfigurable antenna array [132].

An example of a general graph modeling an antenna array is shown in Fig. 6.5, where the connection between vertices D_1 and E_1 represents the constant connection between the two triangular parts of the first array element; while the edge $P_{21}A_2$ represents the connection between one end-point of the switch S_3 and the right lower corner of the second array element as shown in Figs. 6.1 and 6.5. The end-points of each switch are also represented by vertices where an edge between them represents the activation of that switch.

In this chapter we are only interested in studying the complexity of a reconfigurable antenna array. This complexity is directly related to the number of switching elements in the structure and the number of switch-related edges in the corresponding graph model. If a designer wishes to optimize the array's unit cell dimensions as well as the spacing between the different elements, then the complete graph model is needed for that purpose. However, since in this chapter the interest is focused on calculating the complexity of switch reconfigurable antenna arrays, the general graph model can be simplified to only represent the end-points of the various switches while disregarding the other edges [132]. Thus, the graph model shown in Fig. 6.5 can be simplified as shown in Fig. 6.6. The simplified graph, models only the connections between the end-points of the four switches in the two element switch reconfigurable antenna array.

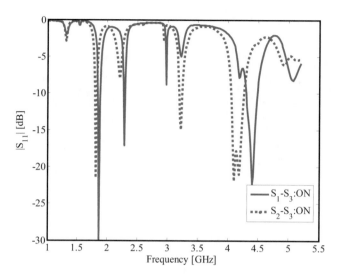

Figure 6.2: The reflection coefficient for the array in Fig. 6.1 for two different switch configurations [132].

Figure 6.3: The fabricated antenna array prototype [132].

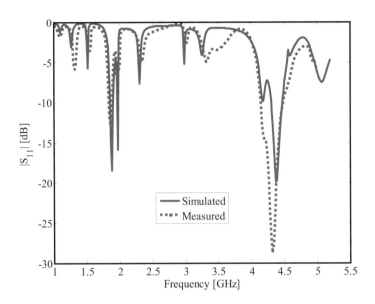

Figure 6.4: A comparison between the measured and simulated reflection coefficient when S1 is activated [132].

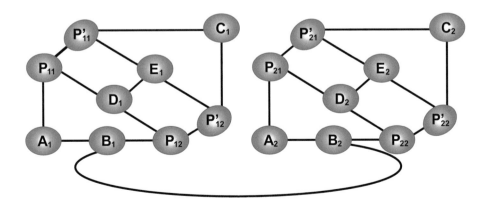

Figure 6.5: General form of a graph model for the antenna array [132].

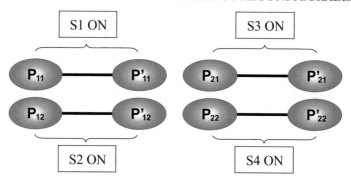

Figure 6.6: Simplified graph model for the array antenna [132].

The graphs representing arrays are frequency dependent [132] and each graph represents a different antenna behavior based on the frequency of operation. This frequency dependence is considered to be on single narrowband frequencies. For example, this array resonates at 2.205 GHz when {S1, S2, S3}, {S2, S3, S4}, or {S3} are ON and at the same time the array preserves the same elliptical polarization for all three cases. Thus, at $f = 2.205$ GHz the array is graph modeled in three graphs {G_1, G_2, G_3} that represent the operation of the antenna at the same frequency while preserving same radiation characteristics. These graphs are considered equivalent. The equivalence is purely based on the antenna array's behavior in different configurations. Thus, equivalent graphs are defined as different models that represent the antenna array in various configurations operating at the same frequency with the same polarization and radiation properties. These graphs shown in Table 6.1 are equivalent at 2.205 GHz, however they are not at other frequencies. Such frequency dependence can be summarized in Eq. (6.2). The use of frequency dependent graphs facilitates the formulation of the antenna's complexity and reliability that are also frequency dependent [117], [132].

$$G_n(f_1) \approx G_m(f_2) \text{ iff } f_1 = f_2 \text{ and same radiation characteristics} \tag{6.2}$$

G_n and G_m model the same antenna array.

6.3   CORRELATION BETWEEN COMPLEXITY AND RELIABILITY

After detailing the graph modeling process of reconfigurable antenna arrays, we discuss the different complexity and reliability parameters governing the reconfigurable antenna array functions. The complexity and reliability for single-element reconfigurable antennas are defined in Chapter 5. In this section, we extend the definitions of single elements to reconfigurable antenna arrays. The complexity of a reconfigurable antenna array is defined for all possible configurations, in all rows and columns of the array.

Table 6.1: Equivalent graphs representing equivalent configurations at 2.205 GHz [132]

6.3.1 THE GENERAL COMPLEXITY OF RECONFIGURABLE ANTENNA ARRAYS

The general complexity of reconfigurable antennas has been previously defined in Chapter 5 as equivalent to the number of edges (NE) in a graph model.

The general complexity of an $M * N$ elements reconfigurable antenna array can be extended from Eq. (5.2) to take into consideration the total number of elements. Hence, the array's general complexity is the summation of each element's general complexity represented as shown in Eq. (6.3).

$$C_{\text{reconfigurable array}} = \sum_{i=1, j=1}^{M,N} NE_{ij} \qquad (6.3)$$

where M represents the number of rows in an array, N the number of columns, NE_{ij} represents the number of edges in the simplified graph (Fig. 6.6) representing all possible configurations in the array elements existing in row i and column j.

The array of Fig. 6.1 is composed of two elements ($M = 1; N = 2$). Its general complexity is:

$$C_{\text{array in Fig. 6.1}} = \sum_{i=1, j=1}^{1,2} NE_{ij} = NE_{11} + NE_{12} = 2 + 2 = 4$$

6.3.2 THE FREQUENCY DEPENDENT COMPLEXITY AND THE CORRELATION WITH RELIABILITY

Based on the fact that a reconfigurable antenna can have different equivalent configurations at the same frequency of operation, a frequency dependent complexity is defined in Chapter 5 for a single-element reconfigurable antenna as shown in Eq. (5.3).

For a reconfigurable antenna array, the frequency dependent complexity is expanded from Eq. (5.3) by taking into account the different elements of the array:

$$C_{\text{reconfigurable array}}(f) = \sum_{i=1, j=1}^{M, N} \underset{K=1, N_{Cij}(f)}{\text{Max}} (NE_K(f)) \tag{6.4}$$

$C_{\text{reconfigurable array}}(f)$ represents the array's complexity at a frequency f

$N_C(f)$ represents the number of equivalent configurations at a frequency f

$NE_{ij}(f)$ represents the number of edges at the configuration i, j for a frequency f.

For example, the frequency-dependent complexity for the array of Fig. 6.1 at 2.205 GHz is according to Table 6.1:

$$C(2.205 \text{ GHz}) = \sum_{i=1, j=1}^{1,2} \underset{K=1,4}{\text{Max}} (NE_K(2.205)) = 2 + 2 = 4$$

The reliability of reconfigurable antennas is derived in Chapter 5 and it depends on the antenna's frequency of operation and its environment as shown in Eq. (5.1).

Extending Eq. (5.1) to a reconfigurable antenna array requires taking into consideration the $M * N$ elements as shown in Eq. (6.5).

$$R_{\text{reconfigurable array}}(f) = \sum_{i,j}^{M, N} \frac{\sum_{k=1}^{N_{Cij}(f)} \sum_{L=1}^{NE_k(f)} P(E_{KL})}{\sum_{k=1}^{N_{Cij}(f)} NE_k(f)} \times 10 \tag{6.5}$$

where:

$R_{\text{reconfigurable array}}(f) = $ The reconfigurable antenna array reliability at a given frequency f

$N_C(f) = $ The number of configurations achieving the frequency f

$NE(f) = $ The number of edges for different configurations at the frequency f

$P(E) = $ Probability of achieving the edge E.

We will now take some examples with actual commercial switches to prove the validity of the derived equations.

The antenna array presented in Fig. 6.1 is assumed to be designed with the p-i-n diodes described in [133]. These p-i-n diodes have an RF switching failure rate of 1.5×10^{-5}(1/h). The probability of failure is then considered to be 0.0015. The probability of achieving an edge in the graph model of Fig. 6.6 is 0.9985 for all switches. The reliability of the array at 2.205 GHz can now be calculated according to Table 6.1 and Eq. (6.5) as follows:

$$
R(2.205) = \sum_{i,j}^{1,2} \frac{\sum_{k=1}^{N_{Cij}(f)} \sum_{L=1}^{NE_k(f)} P(E_{KL})}{\sum_{k=1}^{N_{Cij}(f)} NE_k(2.205)} \times 100 = \frac{\sum_{K=1}^{N_{C11}(1.8)} \sum_{L=1}^{NE_K} P(E_{KL}) + \sum_{K=1}^{N_{C12}(1.8)} \sum_{L=1}^{NE_K} P(E_{KL})}{\sum_{K=1}^{N_{C11}(1.8)} NE_K(2.205) + \sum_{K=1}^{N_{C12}(1.8)} NE_K(2.205)}
$$

$$
= \frac{2 \times 0.9985 + 1 \times 0.9985 + 1 \times 0.9985 + 2 \times 0.9985 + 1 \times 0.9985)}{(2 + 1 + 1 + 2 + 1)} \times 100
$$

$$
= \frac{7 * 0.9985}{(4 + 1 + 1 + 1)}
$$

$$
= 99.85\%
$$

We now incorporate the RF MEMS presented in [134] on the antenna array. Assuming that the probability of switching failure is 0.0357 [134] for the switches in the first element of the array. The probabilities of switching failure are also assumed to be equal to 0.026 for the switches in the second element. Thus, the probability of switching success in the first element is 0.9643 and in the second element it is 0.974. The reliability at 2.205 GHz is then calculated according to Table 6.1 and Eq. (6.5) as follows:

$$
R(2.205) = \sum_{i,j}^{1,2} \frac{\sum_{k=1}^{N_{Cij}(f)} \sum_{L=1}^{NE_k(f)} P(E_{KL})}{\sum_{k=1}^{N_{Cij}(f)} NE_k(2.205)} \times 100 = \frac{\sum_{K=1}^{N_{C11}(1.8)} \sum_{L=1}^{NE_K} P(E_{KL}) + \sum_{K=1}^{N_{C12}(1.8)} \sum_{L=1}^{NE_K} P(E_{KL})}{\sum_{K=1}^{N_{C11}(1.8)} NE_K(2.205) + \sum_{K=1}^{N_{C12}(1.8)} NE_K(2.205)}
$$

$$
= \frac{2 \times 0.9643 + 1 \times 0.9643 + 1 \times 0.974 + 2 \times 0.974 + 1 \times 0.974)}{(2 + 1 + 1 + 2 + 1)} \times 100
$$

$$
= \frac{3 * 0.9643 + 4 * 0.974}{(4 + 1 + 1 + 1)}
$$

$$
= 96.98\%
$$

The reliability of a single element reconfigurable antenna is proven in Chapter 5 in Eq. (5.5) to be correlated with its complexity. Extending this relationship to a reconfigurable antenna array we can say that the reliability of an antenna array at a frequency f can be expressed in terms of its complexity as shown in Eq. (6.6). Using the same expansion as in Chapter 5 (Eq. 5.4), we can also deduce that the reliability is inversely proportional to the complexity of a reconfigurable antenna array at the same frequency f.

$$R_{\text{reconfigurable array}}(f) = \sum_{i,j}^{M,N} \frac{\sum_{k=1}^{N_{C_{ij}}(f)} \sum_{L=1}^{NE_k(f)} P(E_{KL})}{C_{ij}(f) + \sum_{k=1}^{N'_{C_{ij}}(f)} NE_k(f)} \times 100 \tag{6.6}$$

$C(f)$ is calculated in Eq. (6.4)

$N'_C(f)$ is the number of equivalent configurations at a frequency f without the configurations with maximum edges.

6.4   THE CONFIGURATION COMPLEXITY AND THE PRIORITIZATION OF FREQUENCY-DEPENDENT CONFIGURATIONS

In any reconfigurable antenna array, equivalent configurations exist for the same frequency. The failure of any switch affects the complexity and reliability of that particular array. However, the designer is required to prioritize the equivalent frequency configurations based on minimizing the use of switches to improve the overall reliability [132]. For example, in Table 6.1, three different configurations are equivalent at 2.205 GHz. In principle, the primary configuration that achieves this frequency should be the one with the highest reliability.

In previous sections, we have discussed the complexity and reliability of the whole antenna array at a certain frequency of operation. These parameters take into consideration all the different equivalent configurations. One way of prioritizing the array configurations is by calculating the complexity and reliability of each particular configuration independently from the whole array. The complexity of each particular configuration is defined based on Eq. (6.7) as:

$$C_{\text{Config.}}(f) = NE(f) \tag{6.7}$$

where NE is the total number of edges in a particular configuration.

The configuration reliability can be formulated based on Eq. (6.5) and is shown in Eq. (6.8):

$$R_{\text{Config.}}(f) = \frac{\sum_{i=1}^{NE} P(E_i)}{NE(f)} \times 100 = \frac{\sum_{i=1}^{NE} P(E_i)}{C_{\text{config.}}} \times 100 \tag{6.8}$$

Where NE is the total number of edges in a particular configuration.

$P(E)$ is the probability of switching success which is equivalent to the probability of achieving an edge.

To calculate the reliability of the three configurations for the reconfigurable antenna array at 2.205 GHz, let's consider that this array resorts to the RF MEMS from [134] to achieve reconfiguration. The probabilities of switching success are distributed as 0.9643 for switches in Element 1 and 0.974 for switches in Element 2. The reliability of each configuration can be calculated as:

$$R_{G1}(2.205) = \frac{\sum_{i=1}^{NE} P(E_i)}{C_{\text{Config. 1}}} \times 100 = \frac{2 \times 0.9643 + 1 \times 0.974}{3} \times 100$$

$$= 96.75\%$$

$$R_{G2}(2.205) = \frac{\sum_{i=1}^{NE} P(E_i)}{C_{\text{Config. 2}}} \times 100 = \frac{1 \times 0.9643 + 2 \times 0.974}{3} \times 100$$

$$= 97.07\%$$

$$R_{G3}(2.205) = \frac{\sum_{i=1}^{NE} P(E_i)}{C_{\text{Config. 3}}} \times 100 = \frac{1 \times 0.974}{1} \times 100$$

$$= 97.4\%$$

According to the previous results G_3 should be designated as the primary configuration with the highest priority since it has the highest reliability. In the same manner, G_2 and G_1 follow respectively. The configurations represented by G_1 and G_2 are used whenever the configuration represented by G_3 fails. For example, in the case where the switch S3 fails, the configurations G_3 and G_2 cannot be achieved and thus G_1 has to be used with a reliability of 96.75%. This insures the continuous operation of the array at 2.205 GHz [132].

The primary configuration in this example is also the one with the lowest complexity. Lower complexity means a better reliability only when all the probabilities of switching success are equal in the same element; however in an array where the switching success varies between switches, lower complexity won't necessarily mean a better reliability.

For example, if the probability of switching success for S3 was 0.91 while the other probabilities are 0.9643 for {S1, S2} and 0.974 for {S4}. The reliability for G_1 remains the same, however the reliabilities for G_2 and G_3 change and become:

$$R_{G1}(2.205) = \frac{\sum\limits_{i=1}^{NE} P(E_i)}{C_{\text{config.}1}} \times 100 = \frac{2 \times 0.9643 + 1 \times 0.974}{3} \times 100 = 96.75\%$$

$$R_{G2}(2.205) = \frac{\sum\limits_{i=1}^{NE} P(E_i)}{C_{\text{config.}2}} \times 100 = \frac{1 \times 0.9643 + 1 \times 0.974 + 1 \times 0.91}{3} \times 100 = 94.94\%$$

$$R_{G3}(2.205) = \frac{\sum\limits_{i=1}^{NE} P(E_i)}{C_{\text{config.}3}} \times 100 = \frac{1 \times 0.91}{1} \times 100 = 91\%$$

In this case, the primary configuration has to be G_1 since it has the highest reliability and probability of success. Even though it does not exhibit the lowest complexity

$$C_{G1} = 3 > C_{G1} = 1$$

6.5   PRACTICAL DESIGN ASPECTS

Standard operation conditions for common commercialized switches are predefined by the switch manufacturing companies, where temperature, humidity, and other environmental factors are well known, and taken into consideration. Companies clearly specify that they will provide the probability of failure of any of their switching products under unusual conditions to any antenna designer based on specific information and requests from the designer.

For example, the data sheet for the pin diode BAP64-03 by Philips [135] incorporated on the reconfigurable antenna array [136] indicates that the limiting values for this device are in accordance with absolute maximum rating system (IEC 60134). However, the same datasheet clarifies that any stress above one of the normal limiting values causes permanent damage to the device. On the other hand, Ma-Com Corporation, the manufacturer of the pin diode MA4AGCFCP910 used in the reconfigurable antenna array in [137], specifies maximum operating and storing temperatures [138] as well as maximum humidity and moisture tolerance levels for plastic packaged devices [139]. Microsemi's GC 4172 pin diode used in [11] follows the same steps with a comparable temperature limit.

An antenna designer needs to follow five steps to insure better system reliability with less complexity while maintaining versatility and adaptability of the reconfigurable antenna array system. These steps are concluded based on the above mentioned factors, and based on the fact that in unknown and harsh environments, such as in space or on airplanes, standard conditions don't apply and sudden changes in the surrounding conditions are common [132]. The proposed steps are discussed below:

Step 1: identify the environment where the antenna array will be installed and predict possible surrounding abnormalities.

Step 2: Study and customize the switching components with the manufacturing company.

Step 3: Start the antenna design process while maximizing equivalent configurations, graph modeling each configuration, minimizing the number of connections and thus minimizing the number of switches required.

Step 4: Fine tune your design by prioritizing the configurations as described in Section 6.4.

Step 5: Finalize the design to satisfy all the required constraints.

6.6   DISCUSSION

In this chapter, the graph modeling approach of reconfigurable antenna arrays is used to formulate the complexity and reliability of such arrays. These frequency dependent graph models represent different array configurations that are equivalent for any specific frequency. The array's frequency dependent reliability and complexity are proven to be inversely proportional at the same frequency. The rearrangement and prioritization of different equivalent configurations are discussed to improve the antenna performance. The primary configurations at each frequency are defined as the configurations with the highest reliability whether or not they are of the least complexity. These formulations and calculations are useful especially for new applications such as software-defined antennas where the reliability calculations and configurations prioritization are software controlled through a programmable processor. Practical issues are also discussed and steps are proposed to insure a more reliable and less complex reconfigurable antenna array system.

CHAPTER 7

Detection and Correction of Switch Failures in Switch Reconfigurable Antenna Arrays

7.1 INTRODUCTION

After discussing the complexity and reliability in switch reconfigurable antenna arrays in Chapter 6, it is time to consider a method of switch failure detection and propose an efficient correction mechanism. The detection of such failures is a challenging process especially in large arrays where each array element is reconfigured using many switches. Current methods to detect switch failures in these antennas provide poor detection results or require excessive computation time, and are incapable of fixing such failures in a real world application. Recent methods include applications of Case Based Reasoning, Support Vector Machines, and Neural Networks to detect/correct switch failure [140]. However, these methods rely on knowing the radiation pattern characteristics of the antenna array ahead of time. This data is not always available or easily obtained, especially in remote applications such as satellites or harsh environments.

Two detection methods are discussed in this chapter based on 2×1 antenna array with four switches and a 2×2 array with eight switches. The first detects surface current changes by embedding a sensor in each element. The second detects electric field changes by embedding four metallic thin sensing lines in the substrate and measuring their respective coupling behavior. Both detection methods succeed in accurately detecting any switch failure in the reconfigurable antenna array. The two techniques are compared and the best technique is identified based on feasibility and ease of implementation. A correction methodology is also discussed based on using equivalent array configurations that replace defected ones with configurations that restore the antenna radiation characteristics.

7.2 DETECTION OF SWITCH FAILURES IN RECONFIGURABLE ANTENNA ARRAYS

Two switch failure detection techniques are discussed herein. The first technique, sensing points technique, is based on detecting the changes in the array's reflection coefficient taken from several sensing points. The second failure detection technique, sensing lines technique, is based on detecting the variations that the electric fields induce on metallic sensing lines embedded in the antenna substrate.

7.2.1   SENSING POINTS TECHNIQUE

This technique is based on detecting the effect of any failing switch on the surface current distributions of the array elements. Changes in the array's reflection coefficients are detected for each switch failure. However, different switch failures can have similar discrepancies in the array's reflection coefficient response. Thus, the placement of probes in the array elements to act as sensors is proposed. These sensors detect the individual changes in every element and indicate the failure of any particular switch. To prove the concept we use SMA connectors to act as sensors; however a different measurement connector can also be used to satisfy designer constraints. The position of these sensors in every element is critical to prevent its behavior as a sink and thus it is optimized using a computational electromagnetic simulator. This method is broken into four steps as follows [141, 142]:

Step 1: Design an optimized reconfigurable antenna array.

Design a reconfigurable antenna array with the optimized number of antenna elements using the redundancy reduction approach discussed in Chapter 4. The incorporation of switches on each element must result in distinctive and unique responses for each configuration. No redundant switches are present in each element design. The lack of redundant switches ensures the simplest structure of the antenna.

Step 2: Install a sensor in each array element.

Install an additional connector (sensor) in each array element. The position of the sensor is optimized using an EM simulator so that it doesn't interfere with the antenna radiation characteristics. These sensors monitor each element's reflection coefficient. Each switch configuration has a unique reflection coefficient associated with it. The sensed reflection coefficient provides a reference value, which indicates the active switch configuration and identifies a failed switch.

Step 3: Create a lookup table.

Create a lookup table showing all the possible antenna configurations with all possible resonant frequencies. The lookup table also includes the reflection coefficients of each particular array element detected by their respective sensor.

Step 4: Identify failures.

Based on the lookup table created in Step 3, one can identify the normal antenna behavior and operation. Once a change is detected in the array's function, a group of switches that may be causing this change are identified, since a defect may be caused by more than one switch. At this point the sensed reflection coefficients from each element are monitored to identify the specific switch that has failed. The identification of the particular switch is based on the discrepancy that is found between the sensed reflection coefficient in normal operation and the sensed reflection coefficient in defected mode.

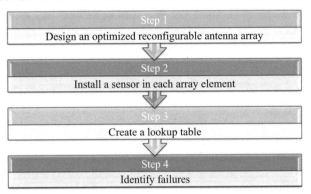

Figure 7.1: Flow chart summarizing the sensing points switch failure detection technique [141, 142].

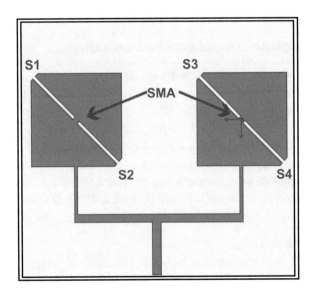

Figure 7.2: Antenna structure with dimensions and sensor positions [141, 142].

A flow chart of this detection methodology is shown in Fig. 7.1.

As an example let's take the two-element antenna array shown in Fig. 7.2. The array is excited through a connector that feeds the different antenna elements through the corporate feeding network. Every array element has an additional connector implanted in the middle of the constant connection dividing the slot. These connectors are used as sensors to identify the failed switches.

Figure 7.3(a) shows the change in the array's reflection coefficient response around 1.866 GHz when S_4 fails. To determine which switch exactly failed, a comparison is achieved between the reflection coefficients measured at normal operation and stored in the lookup table and the ones measured when the change in the array's operation is spotted. These sensed element reflection coefficients are measured from the sensor connectors. A comparison between the sensed reflection coefficient and the one stored in the lookup table is shown in Fig. 7.3(b). The sensed reflection coefficient from element 2 indicates some frequency shifts with respect to normal operation. This frequency shift is distinctive of a failure in switch S_4.

7.2.2   SENSING LINES TECHNIQUE

The second switch failure detection methodology is based on embedding sensing lines in the substrate of a planar antenna array. The sensing lines are thin strips of metal that couple with the electric field generated by the array element. Any change in the antenna configuration alters the footprint that the electric field leaves on these embedded lines. When a switch fails the electric field coupled to these lines changes and thus the failure is detected. This technique can be broken into four steps as shown in Fig. 7.4 [143].

Step 1: Design an optimized reconfigurable antenna array.

Step 2: Embed sensing lines into the array substrate.
    The sensing lines are embedded into the array's substrate in a grid pattern, as shown in Fig. 7.5. Each antenna element has a vertical and horizontal sensing line crossing under it. These lines do not physically intersect and are separated by a layer of foam as shown in Fig. 7.6. The placement of each sensing line is optimized using a simulator. Each sensing line is then assigned a port. To detect a switch failure, the coupling (S_{21}) parameter is measured between the two sensing lines (horizontal and vertical) that cross underneath the element containing the switch in question as shown in Fig. 7.5. For example to detect a failure in Switch 1(S_1) of element 1, S_{21} is measured between the two ports at line 1 (horizontal) and line 3 (vertical) correspondingly.

Step 3: Create a lookup table.
    At the initial design level, the antenna designer collects the S_{21} data for all relevant configurations and stores them in a lookup table.

Step 4: Identify defected switch.
    Once a change in the function is observed a comparison between the measured S_{21} data and the stored one is completed. The failed switch is identified if a discrepancy is observed.
    To illustrate this technique we present the planar 2×2 antenna array shown from a top view in Fig. 7.5. The array is composed of three layers as shown in Fig. 7.6. The top layer is composed of the array elements with the feeding network shown in Fig. 7.5. At the bottom of layer 1, the horizontal sensing lines are present. The dimensions of sensing lines are shown in Table 7.1. The middle section of layer 1 is the Taconic TLY substrate with a dielectric constant of 2.1 and a height of 0.11 cm. Layer 2 is composed of foam with a dielectric constant of 1.0006 and a thickness of 0.35 cm. On top of layer 3 reside the vertical sensing lines. The bottom of layer 3 is composed of a

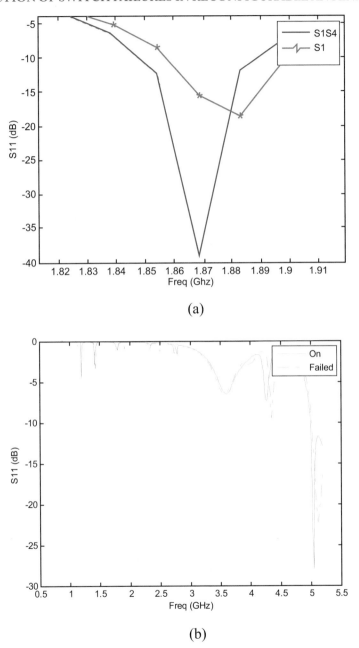

Figure 7.3: (a) A comparison of the reflection coefficient parameter for the normal operation ($S_1 S_4$ON), with one failed switch. (b) The S_{11} value measured at the sensor in the second element showing a frequency shift indicating a failure in the switch S_4 [141, 142].

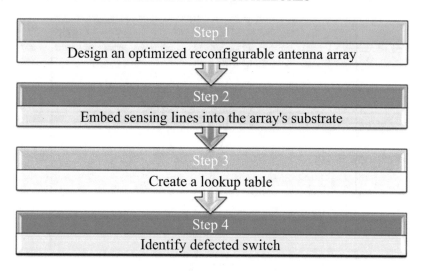

Figure 7.4: Flow chart of the second failure detection methodology [143].

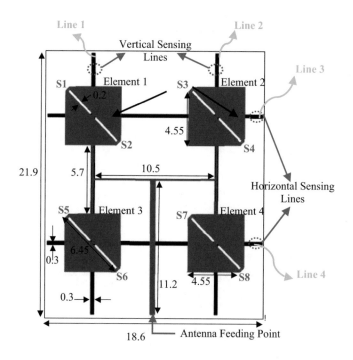

Figure 7.5: Planar antenna array with dimensions and the sensing lines grid pattern [143].

common ground plane. The middle section of layer 3 is the substrate Taconic TLY with a dielectric constant of 2.1 and a thickness of 0.11 cm. The foam layer (layer 2) is very essential since it isolates layer 1 from layer 3 and thus prevents the horizontal and vertical sensing lines from touching. The fabricated prototype is shown in Fig. 7.7(a) and the horizontal sensing layer at the bottom of layer 1 is shown in Fig. 7.7(b).

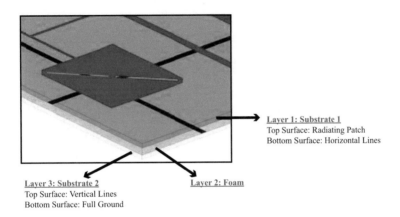

Layer 1: Substrate 1
Top Surface: Radiating Patch
Bottom Surface: Horizontal Lines

Layer 3: Substrate 2
Top Surface: Vertical Lines
Bottom Surface: Full Ground

Layer 2: Foam

Figure 7.6: Layering of the antenna array structure where foam and sensing lines are sandwiched between the two layers of substrate [143].

Table 7.1: Dimensions of Sensing Lines [141]

	Horizontal Lines	Vertical Lines
Elevation	0.46 cm	0.11 cm
Length	18.6 cm	21.9 cm
Width	0.3 cm	0.3 cm

To detect a failure in switch $1(S_1)$ of element 1, S_{21} is measured between the two ports at line 1 and line 3, correspondingly. A comparison between the measured and simulated S_{21} (line 1, line 3) is shown in Fig. 7.8 when all switches are ON. This detection is based on comparing the $|S_{21}|$ of the appropriate sensing lines with the stored ones once a change in the array's function is observed. Fig. 7.9 shows a comparison between the S_{21} parameters (line 1, line 3) when all switches are ON

Figure 7.7: (a) The fabricated prototype with the different ports, (b) The horizontal sensing lines [143].

and when S_1 fails. A change in S_{21} is noticed, the change corresponds to the failure of switch S_1. This proves the validity of this approach in accurately identifying failed switches [143].

7.3   COMPARISON BETWEEN THE TWO DETECTION TECHNIQUES

The first detection technique proposed in Section 7.2.1 integrates sensors into each element to gather centralized data from each element of the reconfigurable antenna array. Some practical problems arise from the implementation of this technique. The first problem is directly related to the number of connectors needed to accomplish the failure detection. As the number of elements gets large, the number of connections to monitor becomes excessive. For example, in a 100×100 elements antenna array 10,000 sensors must be integrated into the array antenna. This is an excessive number of connectors that quickly becomes impractical to monitor.

The second problem resides in the weight of the whole array resulting from the addition of sensors. As the number of these connectors becomes large, the weight and physical size for the antenna array add up.

These problems make the first detection methodology convenient for antenna arrays of small number of elements; however this technique is proven to be impractical on large antenna arrays.

All the issues that are found with the first failure detection are efficiently solved by the second detection methodology. The sensing lines technique provides several advantages over the

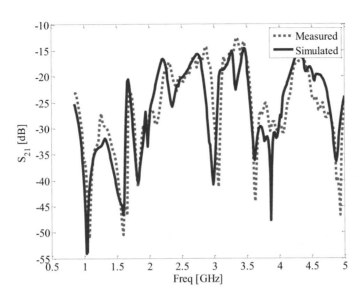

Figure 7.8: A comparison between the measured and simulated S_{21} parameter when all switches are ON for lines 1 and 3 [143].

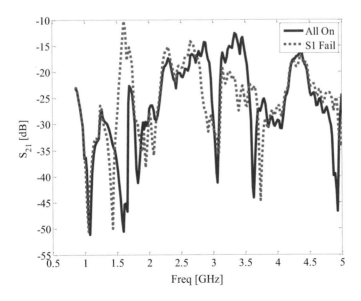

Figure 7.9: A comparison of S_{21} data when all switches are ON and when S_1 fails [143].

first technique and does not resort to integrated sensors in each element. The second technique proposes embedding thin metallic sensing lines into the substrate of the planar antenna array. The embedding of these sensing lines vastly reduces the number of evaluation ports and improves the efficiency of the detection mechanism. The number of evaluation ports saved with the second failure detection technique is expressed by Eq. (7.1).

$$\text{Ports Saved} = (M \times N) - (M + N) \tag{7.1}$$

where M and N represent the number of elements in a $M \times N$ antenna array.

Table 7.2 shows the number of ports saved by the sensing lines failure detection technique in reconfigurable antenna arrays.

Table 7.2: Number of ports saved by the second technique as the number of array elements increases [141]

Antenna Size:	# of Elements:	# of Ports:	Ports Saved:
2x2	4	4	0
2x3	6	5	1
9x9	81	18	63
100x100	10,000	200	9,800

7.4 OVERCOMING SWITCH FAILURES IN RECONFIGURABLE ANTENNA ARRAYS USING FREQUENCY-DEPENDENT GRAPHS

To overcome switch failures in reconfigurable antenna arrays, these failures need to be detected and identified by following one of the detection methodologies discussed previously. Equivalent array configurations can be used to bypass the defected switches. Equivalent configurations are configurations that give the same frequency response while preserving the same radiation characteristics. Table 7.3 presents the equivalent antenna configurations of the antenna array shown in Fig. 7.2, for $f = 1.866$ GHz, 2.29 GHz, 4.335 GHz, and 4.445 GHz. The detection of any switch failure such as S_4 can be done using any of the detection techniques discussed previously. For example, the failure of switch (S_4) eliminates the resonance at $f = 1.866$; however Table 7.3 indicates that when (S_1, S_3) or (S_2, S_3) are ON, the frequency $f = 1.866$ GHz is restored as shown in Fig. 7.10. Figure 7.11(a) shows the array's radiation pattern for the configuration $(S_1 S_4)$ ON. The radiation pattern also exhibits a noticeable shift in direction due to the failure of S_4 as shown in Fig. 7.11(b). Once the correcting configuration $S_1 S_3$ is ON and the failed switch S_4 is bypassed, the array pattern shifts back to normal operation as shown in Fig. 7.11(c) which proves that the correction methodology does not affect the antenna's radiation characteristics.

Table 7.3: Lookup table with equivalent frequencies for the array in Fig. 7.2

Frequency	Equivalent Configurations				
F = 1.866 GHz	$S_1 S_4$	$S_1 S_3$	$S_2 S_3$	S_4	
F = 2.29 GHz	$S_1 S_3$	$S_1 S_4$	$S_2 S_3$		
F = 4.335 GHz	OFF	$S_2 S_3$	$S_1 S_3$	S_3	$S_1 S_4$
F = 4.445 GHz	$S_1 S_4$	S_1	$S_1 S_3$	S_3	

Graph modeling the array's configurations allows the switch failure identification and the search for a correcting configuration to be software controlled and more efficient. Whenever a switch failure is identified, the lost operating frequency is identified and the configuration using the defected switch is eliminated. The use of frequency-dependent graphs allows the use of their search algorithms to achieve a very fast and swift action to overcome failures, and avoid the defected switches by utilizing the equivalent configurations [144].

The same correction mechanism is also applied to the array antenna shown in Fig. 7.5. As-suming that the switch S_4 fails, the array operation at 3.65 GHz is lost as shown in Fig. 7.12. Once the lookup table is referenced the configuration is restored with the elimination of S_4 and activation of S_2 and S_3 simultaneously as shown in Fig. 7.12.

7.5   DISCUSSION

In this chapter we have presented two switch failure detection techniques. These techniques are applied on reconfigurable antenna arrays to identify failed switching components. The first technique (sensing points technique) is based on implanting sensors in strategic positions in every element of the antenna array. These sensors measure the reflection coefficient for each element under normal

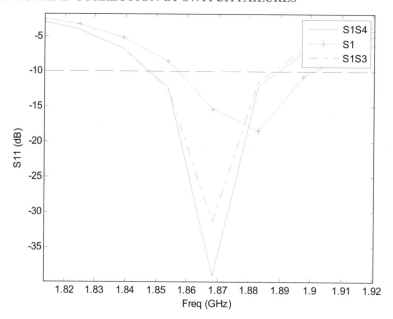

Figure 7.10: Overcoming of S_4 failure by using the equivalent configuration $S_1 S_3$ [144].

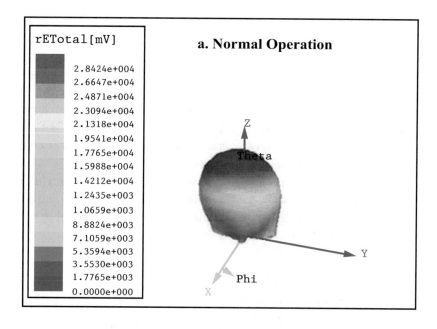

Figure 7.11: (a) The normal operation radiation pattern of the array in Fig. 7.2 at 1.866 GHz [142] *Continues.*

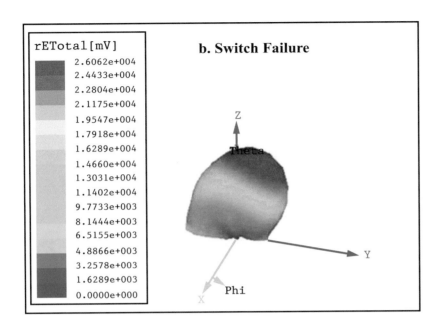

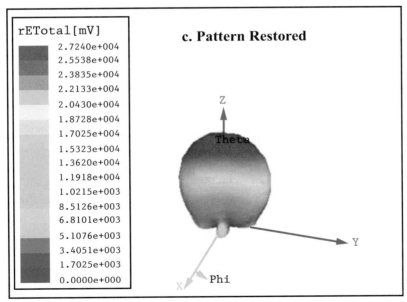

Figure 7.11: *Continued.* (b) The radiation pattern once the switch S_4 fails, (c) The radiation restored with the $S_1 S_3$ configuration [142].

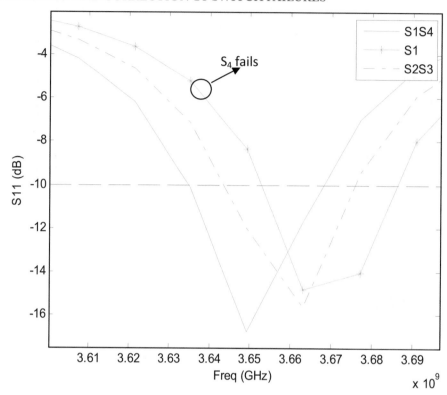

Figure 7.12: Overcoming the failure in S_4 by using the $S_2 S_3$ configuration [141].

operation and store the data in a lookup table. Once a failure is suspected, these sensors execute the same measurement again and compare their results with the stored "normal" data and identify which switch exactly has failed.

The second technique (sensing lines techniques) is based on measuring the coupling between sensing lines that are incorporated into the array's structure. These sensing lines intersect under each array element separated by a layer of foam to avoid physical contact. Once a failure is suspected the coupling is measured between the respective lines and the results are compared to the coupling measured under normal operation and stored in a lookup table.

Overcoming these failures is based on equivalent configurations, where configurations that don't resort to the defected switch are used to restore the frequency of operation as well as the radiation pattern to its normal operation. The whole process can be automated and software controlled with fast searching algorithms using the graph models of the respective array's configurations.

CHAPTER 8

Conclusion

In this book we provide the antenna designer with a new, easy-to-implement plan of action for reconfigurable antenna design. The implementation of reconfigurable antennas on new wireless communication platforms has increased the demand for self-automated antennas with high tolerance for failure. These antenna systems that are required to change their functions according to evolving circumstances need to be non-redundant, reliable, and able to overcome any failure. We try to provide the designer with answers to important questions. These questions are discussed in Chapter 1, and allow the designer to choose which reconfigurable property to modify, how to arrange the antenna topology, and which reconfiguration technique to use.

In Chapter 2, we propose rules to graph model reconfigurable antennas. These rules are not unique and many other rules can be created to graph model reconfigurable antennas. However, the rules that are proposed in Chapter 2 are the basis for the analysis executed in the rest of the book. Designers modeling their antennas using these rules should get to the same results discussed herein.

An iterative design approach is discussed in Chapter 3. This approach is designed to generate an optimal antenna structure that doesn't include any redundant components, while maintaining the same performance. This technique is the basis for a redundancy reduction method that is discussed in Chapter 4. The redundancy reduction approach is based on considering each unique path in a graph responsible for a distinctive function. By removing redundant components from the antenna structure the designer is reducing the cost of design as well as reducing numerous losses in the anticipated system. The redundancy reduction approach aims at reducing the time needed for designing reconfigurable antenna structures by mathematically indicating the number of required elements.

The complexity and reliability of reconfigurable antennas and reconfigurable antenna arrays are discussed in Chapters 5 and 6 respectively. Continuous operation of non-redundant antenna structures under unknown environments and harsh conditions is proposed. Detecting and overcoming switch failures in reconfigurable antenna arrays is proposed in Chapter 7 where the whole process can be software controlled and automated.

Finally, in this book we have discussed graph models as new tools for the modeling, redundancy reduction, and controlling of reconfigurable antennas. Graph models transform reconfigurable antennas from their bulky mechanical states into software accessible devices that can be optimized, trained, corrected, and searched for efficient communication purposes. It is a new method proposed herein to allow the antenna designer to mathematically and efficiently reshape antenna topologies for better implementation in real life applications.

Bibliography

[1] D. Schaubert, "Frequency-agile polarization diversity microstrip antennas and frequency scanned arrays," US Patent #4,367,474, Jan. 1983. 1

[2] J. K. Smith, "Reconfigurable aperture antenna (RECAP)," DARPA, 1999. 1

[3] C. G. Chrisotdoulou, Y. Tawk, S. A. Lane, and S. R. Erwin, "Reconfigurable Antennas for Wireless and Space Applications," Proceedings of the IEEE, 100(7), pp. 2250–2261, 2012. DOI: 10.1109/JPROC.2012.2188249. 1, 2, 3, 8

[4] Y. Tawk, "Analysis, Design and Implementation of Front-End Reconfigurable Antenna Systems (FERAS)," Ph.D. Dissertation, 2011. 1

[5] C. W. Jung, M. Lee, G. P. Li, and F. De Flaviis, "Reconfigurable scan-beam single-arm spiral antenna integrated with RF-MEMS switches," IEEE Transactions on Antennas and Propagation, vol. 54, no. 2, pp. 455–463, Feb. 2006. DOI: 10.1109/TAP.2005.863407. 1, 4

[6] G. H. Huff and J. T. Bernhard, "Integration of packaged RF-MEMS switches with radiation pattern reconfigurable square spiral microstrip antennas," IEEE Transactions on Antennas and Propagation, vol. 54, no. 2, pp. 464–469, Feb. 2006. DOI: 10.1109/TAP.2005.863409. 75, 91

[7] B. A. Cetiner, G. R. Crusats, L. Jofre, and N. Biyikli, "RF MEMS integrated frequency reconfigurable annular slot antenna," IEEE Transactions on Antennas and Propagation, vol. 58, no. 3, pp. 626–632, Mar. 2010. DOI: 10.1109/TAP.2009.2039300.

[8] A. Grau, J. Romeu, M. Lee, S. Blanch, L. Jofre, and F. De Flaviis, "A dual linearly polarized MEMS-reconfigurable antenna for narrowband MIMO communication systems," IEEE Transactions on Antennas and Propagation, vol. 58, no. 1, pp. 4–16, Jan. 2010. DOI: 10.1109/TAP.2009.2036197.

[9] S. Nikolaou, N. D. Kingsley, G. E. Ponchak, J. Papapolymerou, and M. M. Tentzeris, "UWB elliptical monopoles with a reconfigurable band notch using MEMS switches actuated without bias lines," IEEE Transactions on Antennas and Propagation, vol. 57, no. 8, pp. 2242–2251, Aug. 2009. DOI: 10.1109/TAP.2009.2024450.

[10] E. Erdil, K. Topalli, M. Unlu, O. A. Civi, and T. Akin, "Frequency tunable patch antenna using RF MEMS technology," IEEE Transactions on Antennas and Propagation, vol. 55, no. 4, pp. 1193–1196, Apr. 2007. DOI: 10.1109/TAP.2007.893426. 1

[11] S. Shelley, J. Costantine, C. G. Christodoulou, D. E. Anagnostou, and J. C. Lyke, "FPGA-controlled switch-reconfigured antenna," IEEE Antennas and Wireless Propagation Letters, vol. 9, pp. 355–358, 2010. DOI: 10.1109/LAWP.2010.2048550. 1, 4, 6, 59, 102

[12] D. Peroulis, K. Sarabandi, and L. P. B. Katehi, "Design of reconfigurable slot antennas," IEEE Transactions on Antennas and Propagation, vol. 53, no. 2, pp. 645–654, Feb. 2005. DOI: 10.1109/TAP.2004.841339.

[13] M. K. Fries, M. Grani, and R. Vahldieck, "A reconfigurable slot antenna with switchable polarization," IEEE Microwave and Wireless Components Letters, vol. 13, no. 11, pp. 490–492, Nov. 2003. DOI: 10.1109/LMWC.2003.817148.

[14] S. Nikalaou, R. Bairavasubramanian, C. Lugo, I. Carrasquillo, D. C. Thompson, G. E. Ponchak, J. Papapolymerou, and M. M. Tentzeris, "Pattern and frequency reconfigurable annular slot using PIN diodes," IEEE Transactions on Antennas And Propagation, vol. 54, no. 2, pp. 439–447, Feb. 2006. DOI: 10.1109/TAP.2005.863398.

[15] N. Jin, F. Yang, and Y. Rahmat-Samii, "A novel patch antenna with switchable slot (PASS): dual-frequency operation with reversed circular polarizations," IEEE Transactions on Antennas and Propagation, vol. 54, no. 4, pp. 1031–1043, Mar. 2006. DOI: 10.1109/TAP.2006.869939.

[16] S.-H. Chen, J.-S. Row, and K.-L Wong, "Reconfigurable square-ring patch antenna with pattern diversity," IEEE Transactions on Antennas and Propagation, vol. 55, no. 2, pp. 472–475, Feb. 2007. DOI: 10.1109/TAP.2006.889950.

[17] S.-J. Wu and T.-G. Ma, "A wideband slotted bow-tie antenna with reconfigurable CPW-to slotline transition for pattern diversity," IEEE Transactions on Antennas and Propagation, vol. 56, no. 2, pp. 327–334, Feb. 2008. DOI: 10.1109/TAP.2007.915454.

[18] B. Kim, B. Pan, S. Nikolaou, Y.-S Kim, J. Papapolymerou, and M. M. Tentzeris, "A novel single-feed circular microstrip antenna with reconfigurable polarization capability," IEEE Transactions on Antennas and Propagation, vol. 56, no. 3, pp. 630–638, Mar. 2008. DOI: 10.1109/TAP.2008.916894.

[19] R.-H Chen and J.-S Row, "Sing-fed microstrip patch antenna with switchable polarization," IEEE Transactions on Antennas and Propagation, vol. 56, no. 4, pp. 922–926, Apr. 2008. DOI: 10.1109/TAP.2008.919211.

[20] J. Sarrazin, Y. Mahe, S. Avrillon, and S. Toutain, "Pattern reconfigurable cubic antenna," IEEE Transactions on Antennas and Propagation, vol. 57, no. 2, pp. 310–317, Feb. 2009. DOI: 10.1109/TAP.2008.2011221.

[21] J. Perruisseau-Carrier, P. Pardo-Carrera, and P. Miskovsky, "Modeling, design and characterization of a very wideband slot antenna with reconfigurable band rejection," IEEE

Transactions on Antennas and Propagation, vol. 58, no. 7, pp. 2218–2226, Jul. 2010. DOI: 10.1109/TAP.2010.2048872.

[22] P.-Y. Qin, A. R. Weily, Y. J. Guo, T. S. Bird, and C.-H. Liang, "Frequency reconfigurable quasi-yagi folded dipole antenna," IEEE Transactions on Antennas and Propagation, vol. 58, no. 8, pp. 2742–2747, Aug. 2010. DOI: 10.1109/TAP.2010.2050455. 1

[23] N. Behdad and K. Sarabandi, "A varactor-tuned dual-band slot antenna," IEEE Transactions on Antennas and Propagation, vol. 54, no. 2, pp. 401–408, Feb. 2006. DOI: 10.1109/TAP.2005.863373. 2

[24] E. A.-Daviu, M. C.-Fabres, M. F.-Bataller, and A. V.-Jimenez, "Active UWB antenna with tunable band-notched behavior," IEEE Electronics Letters, vol. 43, no. 18, pp. 959–960, Aug. 2007. DOI: 10.1049/el:20071567.

[25] W.-S. Jeong, S.-Y. Lee, W.-G. Lim, H. Lim, and J.-W. Yu, "Tunable band-notched ultra wideband (UWB) planar monopole antennas using varactor," 38th European Microwave Conference, Oct. 2008, pp. 266–268. DOI: 10.1109/EUMC.2008.4751439.

[26] C. R. White and G. M. Rebeiz, "Single and dual-polarized tunable slot-ring antennas," IEEE Transactions on Antennas and Propagation, vol. 57, no. 1, pp. 19–26, Jan. 2009. DOI: 10.1109/TAP.2008.2009664.

[27] H. Jiang, M. Patterson, C. Zhang, and G. Subramanyan, "Frequency tunable microstrip patch antenna using ferroelectric thin film varactor," IEEE National Aerospace & Electronics Conference, pp. 248–250, Jul. 2009. DOI: 10.1109/NAECON.2009.5426620.

[28] S.-S. Oh, Y.-B. Jung, Y. –R. Ju, and H.-D. Park, "Frequency-tunable open ring microstrip antenna using varactor," International Conference on Electromagnetics in Advanced Applications, pp. 624–626, Sept. 2010. DOI: 10.1109/ICEAA.2010.5652325.

[29] S. L-S. Yang, A. A. Kishk, and Lee Kai-Fong, "Frequency reconfigurable U-slot microstrip patch antenna," IEEE Antennas and Wireless Propagation letters, vol. 7, pp. 127–129, Jan. 2008. DOI: 10.1109/LAWP.2008.921330. 2

[30] L. N. Pringle, P. H. Harms, S. P. Blalock, G. N. Kiesel, E. J. Kuster, P. G. Friederich, R. J. Prado, J. M. Morris, and G. S. Smith, "A reconfigurable aperture antenna based on switched links between electrically small metallic patches," IEEE Transactions on Antennas and Propagation, vol. 52, no. 6, pp. 1434–1445, Jun. 2004. DOI: 10.1109/TAP.2004.825648. 2

[31] C. J. Panagamuwa, A. Chauraya, and J. C. Vardaxoglou, "Frequency and beam reconfigurable antenna using photoconductive switches," IEEE Transactions on Antennas and Propagation, vol. 54, no. 2, pp. 449–454, Feb. 2006. DOI: 10.1109/TAP.2005.863393.

[32] M. R. Chaharmir, J. Shaker, M. Cuhaci, and A.-R. Sebak, "Novel photonically-controlled reflectarray antenna," IEEE Transactions on Antennas and Propagation, vol. 54, no. 4, pp. 1134–1141, Apr. 2006. DOI: 10.1109/TAP.2006.872644.

[33] Y. Tawk, A. R. Albrecht, S. Hemmady, G. Balakrishnan, and C. G. Christodoulou, "Optically pumped frequency reconfigurable antenna design," IEEE Antennas and Wireless Propagation Letters, vol. 9, pp. 280–283, Mar. 2010. DOI: 10.1109/LAWP.2010.2047373.

[34] Y. Tawk, J. Costantine, S. E. Barbin, and C. G. Christodoulou, "Integrating laser diodes in a reconfigurable antenna system," SBMO/IEEE MTT-S International Microwave and Optoelectronics Conference, Oct. 2011. DOI: 10.1109/IMOC.2011.6169295. 2

[35] G. M. Rebeiz and J. B. Muldavin, "RF-MEMS Switches and Switch Circuits," IEEE Microwave Magazine, vol. 2, no. 4, pp. 59–71, Dec. 2001. DOI: 10.1109/6668.969936. 2

[36] U. L. Rohde and D. P. Newkirk, "RF/microwave circuit design for wireless applications," John Wiley & Sons, 2000. DOI: 10.1002/0471224138.

[37] I. Gutierrez, E. Hernandez, and E. Melendez, "Design and characterization of integrated varactors for RF applications," John Wiley & Sons, 2006. DOI: 10.1002/9780470035924.

[38] C. Kittel, "Introduction to solid state physics," John Wiley & Sons, Seventh Edition, 1996. 2

[39] S. Jalali Mazlouman, M. Soleimani, A. Mahanfar, C. Menon, and R. G. Vaughan, "Pattern reconfigurable square ring patch antenna actuated by hemispherical dielectric elastomer," Electronics Letters, vol. 47, no. 3, pp. 164–165, Feb. 2011. DOI: 10.1049/el.2010.3585. 2

[40] J.-C Langer, J. Zou, C. Liu, and J. T. Bernhard, "Reconfigurable out-of-plane microstrip patch antenna using MEMS plastic deformation magnetic actuation," IEEE Microwave and Wireless Components Letters, vol. 13, no. 3, pp. 120–122, Mar. 2003. DOI: 10.1109/LMWC.2003.810123.

[41] Y. Tawk and C. G. Christodoulou, "A cellular automata reconfigurable microstrip antenna design," IEEE International Symposium on Antennas and Propagation, Jun. 2009, pp. 1–4. DOI: 10.1109/APS.2009.5171548.

[42] Y. Tawk, J. Costantine, and C. G. Christodoulou, "A frequency reconfigurable rotatable microstrip antenna design," IEEE International Symposium on Antennas and Propagation, Jul. 2010, pp. 1–4. DOI: 10.1109/APS.2010.5561272. 2

[43] W. Hu, M. Y. Ismail, R. Cahill, J. A. Encinar, V. Fusco, H. S. Gamble, D. Linton, R. Dickie, N. Grant, and S. P. Rea, "Liquid–crystal-based reflectarray antenna with electronically switchable monopulse patterns," Electronics Letters, vol. 43, no. 14, Jul. 2007. DOI: 10.1109/LMWC.2003.810123. 2

[44] L. Liu and R. J. Langley, "Liquid crystal tunable microstrip patch antenna," Electronics Letters, vol. 44, no. 20, pp. 1179–1180, Sept. 2008. DOI: 10.1049/el:20081995.

[45] D. M. Pozar and V. Sanchez, "Magnetic tuning of a microstrip antenna on a ferrite substrate," Electronics Letters, vol. 24, no. 12, pp. 729–731, Jun. 1988. DOI: 10.1049/el:19880491.

[46] L. Dixit and P.K.S. Pourush, "Radiation characteristics of switchable ferrite microstrip array antenna," IEE Proceedings Microwaves, Antennas and Propagation, vol. 147, no. 2, pp. 151–155, Apr. 2000. DOI: 10.1049/ip-map:20000038. 2

[47] J. Costantine, "Design, optimization and analysis of reconfigurable antennas," Ph.D. Dissertation, Dec. 2009. 2, 6, 14, 15, 16, 40, 41, 42, 55, 76

[48] Y. Tawk, J. Costantine, and C. G. Christodoulou, "A varactor based reconfigurable filtenna," IEEE Antennas and Wireless Propagation, vol. 11, pp. 716–719, 2012. DOI: 10.1109/LAWP.2012.2204850. 5, 7

[49] FCC, "Report of the spectrum efficiency working group," FCC spectrum policy task force, Tech. Rep., Nov. 2002. 5

[50] J. Mitola, "Cognitive Radio: An integrated agent architecture for software defined radio," Ph.D. dissertation, 2000. 5

[51] Y. Tawk, J. Costantine, K. Avery, and C. G. Christodoulou, "Implementation of a cognitive radio front-end using rotatable controlled reconfigurable antennas," IEEE Transactions on Antennas and Propagation, vol. 59, no. 5, pp. 1773–1778, May 2011. DOI: 10.1109/TAP.2011.2122239. 5

[52] Y. Tawk and C. G. Christodoulou, "A new reconfigurable antenna design for cognitive radio," IEEE Antennas and Wireless Propagation Letters, vol. 8, pp. 1378–1381, 2009. DOI: 10.1109/LAWP.2009.2039461.

[53] Y. Tawk, M. Bkassiny, G. El-Howayek, S. K. Jayaweera, K. Avery, and C. G. Christodoulou, "Reconfigurable front-end antennas for cognitive radio applications," IET Microwaves, Antennas & Propagation, vol. 5, no. 8, pp. 985–992, Jun. 2011. DOI: 10.1049/iet-map.2010.0358.

[54] E. Ebrahimi, J. R. Kelly, and P. S. Hall, "Integrated wide-narrowband antenna for multi-standard radio," IEEE Transactions on Antennas and Propagation, vol. 59, no. 7, pp. 2628–2635, Jul. 2011. DOI: 10.1109/TAP.2011.2152353.

[55] G. T. Wu, R. L. Li, S. Y. Eom, S. S. Myoung, K. Lim, J. Laskar, S. I. Jeon, and M. M. Tentzeris, "Switchable quad-band antennas for cognitive radio base station applications," IEEE Transactions on Antennas and Propagation, vol. 58, no. 5, pp. 14668–1476, May 2010. DOI: 10.1109/TAP.2010.2044472. 5

[56] D. Piazza, N. J. Kirsch, A. Forenza, R. W. Heath, and K. R. Dandekar, "Design and evaluation of a reconfigurable antenna array for MIMO systems," IEEE Transactions on Antennas and Propagation, vol. 56, no. 3, pp. 869–881, Mar. 2008. DOI: 10.1109/TAP.2008.916908. 8, 91, 92

[57] B. A. Cetiner, H. Jafarkhani, J.–Y. Qian, H. J. Yoo, A. Grau, and F. De Flaviis, "Multifunctional reconfigurable MEMS integrated antennas for adaptive MIMO systems," IEEE Communication Magazine, vol. 42, no. 12, pp. 62–70, Dec. 2004. DOI: 10.1109/MCOM.2004.1367557.

[58] Z. Li, Z. Du, and K. Gong, "Compact reconfigurable antenna array for adaptive MIMO systems," IEEE Antennas and Wireless Propagation Letters, vol. 8, pp. 1317–1321, 2009. DOI: 10.1109/LAWP.2009.2038182.

[59] P. –Y Qin. Y. Jay Guo, and C.-H Liang, "Effect of antenna polarization diversity on MIMO system capacity," IEEE Antennas and Wireless Propagation Letters, vol. 9, pp. 1092–1095, 2010. DOI: 10.1109/LAWP.2010.2093116. 8

[60] Y. Tawk, J. Costantine, and C. G. Christodoulou, "A Reconfigurable Bans- Reject MIMO Antenna for Cognitive Radio," 7th European Conference on Antennas and Propagation, 2013. 8

[61] R. Mizzoni et al., "Ku-band telecommunication antennas with limited pattern reconfiguration," 28th European Space Agency Workshop, Jun. 2005, pp. 651–656. 9

[62] A. G. Roederer, "Antennas for space: some recent European developments and trends," 18th International Conference on Applied Electromagnetics and Communications, Oct. 2005. DOI: 10.1109/ICECOM.2005.204908.

[63] M. Ali, A. T. M. Sayem, and V. K. Kunda, "A Reconfigurable Stacked Microstrip Patch Antenna for Satellite and Terrestrial Links," IEEE Transactions on Vehicular Technology, vol. 56, no. 2, pp. 426–435, Mar. 2007. DOI: 10.1109/TVT.2007.891412.

[64] S. Rao, Minh, and C.-Chien Hsu, "Reconfigurable antenna system for satellite communications," IEEE International Symposium on Antennas and Propagation, Jun. 2007, pp. 3157–3160. DOI: 10.1109/APS.2007.4396206.

[65] H. Liu, S. Gao, and T. H. Loh, "Circularly polarized electronically steerable parasitic array radiator antenna for satellite," Fourth European Conference on Antennas and Propagation, Apr. 2010, pp. 1–4. DOI: 10.1155/2010/213576. 9

[66] T. H. Cormen, C. E. Leiserson, R. L. Rivest, and C. Stein, "Introductions to Algorithms," MIT press, 2001. 11, 14, 27, 30

[67] E. Klavins, "Programmable Self Assembly," IEEE Control Systems Magazine, vol. 27, no. 4, pp. 43–56, Aug. 2007. DOI: 10.1109/MCS.2007.384126. 11

[68] E. Klavins, "Programmable Self Assembly," IEEE Control Systems Magazine, vol. 27, no. 4, pp. 43–56, Aug. 2007. DOI: 10.1109/MCS.2007.384126. 11

[69] E. Klavins, R. Ghrist, and D. Lipsky, "Graph Grammars for Self Assembling Robotic Systems," IEEE International Conference on Robotics and Automation, vol. 5, pp. 5293–5300, April 2004. DOI: 10.1109/ROBOT.2004.1302558.

[70] N. Napp, S. Burden, and E. Klavins, "The Statistical Dynamics of Programmed Assembly," IEEE International Conference on Robotics and Automation, pp. 1469–1476, May 2006. DOI: 10.1109/ROBOT.2006.1641916.

[71] E. Klavins, "Self-Assembly From the point of view of its pieces," American Control Conference, pp. 7, June 2006. DOI: 10.1109/ACC.2006.1655325. 11

[72] J. Clark and Derek Alan Holton, "A First Look At Graph Theory," World Scientific Publishing Company, 1991. DOI: 10.1142/1280. 12, 14, 15, 52

[73] R. Diestel, "Graduate Texts in Mathematics: Graph Theory," Springer, 2000.

[74] J. A. Bondy and U. S. R. Murty, "Graduate Texts in Mathematics: Graph Theory," Springer, 2008. DOI: 10.1007/978-1-84628-970-5.

[75] J. M. Aldous and R. J. Wilson, "Graphs and Applications: An Introductory Approach," vol. 1, Springer Verlag, 2000. DOI: 10.1007/978-1-4471-0467-4.

[76] R. J. Wilson and J. J. Watkins, "Graphs: An introductory approach," John Wiley and Sons, 1990.

[77] G. Chartrand, "Introductory Graph Theory," unabridged edition, Dover Publications, 1984. 12, 52

[78] A. Patnaik, D. E. Anagnostou, C. G. Christodoulou and J. C. Lyke, "Neurocomputational analysis of a multiband reconfigurable planar antenna," IEEE Transactions on Antennas and Propagation, vol. 53, Issue 11, pp. 3453–3458, Nov. 2005. DOI: 10.1109/TAP.2005.858617. 16, 17, 18

[79] A. E. Fathy, A. Rosen, H. S. Owen, F. McGuinty, D. J. Mcgee, G. C. Taylor, R. Amantea , P. K. Swain, S. M. Perlow and M. Elsharbiny, "Silicon-based reconfigurable antennas-concepts, analysis, implementation and feasibility," IEEE Transactions on Microwave Theory and Techniques, vol. 51, Issue 6, pp. 1650–1661, June 2003. DOI: 10.1109/TMTT.2003.812559. 17, 19

[80] L. M. Feldner, C. D. Nordquist, and C. G. Christodoulou, "RFMEMS Reconfigurable Triangular Patch Antenna," IEEE AP/URSI International Symposium, vol. 2A, pp. 388–391, July 2005. DOI: 10.1109/APS.2005.1551824. 20, 71, 72

[81] C. S. Deluccia, D. H. Werner, P. L. Werner, M. Fernandez Pentoja, and A. R. Bretones, "A novel frequency agile beam scanning reconfigurable antenna," IEEE Antennas and Propagation society international symposium, vol. 2, pp. 1839–1842, June 2004. DOI: 10.1109/APS.2004.1330558. 21, 22

[82] N. Behdad and K. Sarabandi, "Dual-Band reconfigurable antenna with a very wide tunability range," IEEE Transactions on Antennas and Propagation, vol. 54, Issue 2, Part 1, pp. 409–416, Feb. 2006. DOI: 10.1109/TAP.2005.863412. 23, 24

[83] J. C. Langer, J. Zou, C. Liu, and J. T. Bernhard, "Micromachined Reconfigurable Out Of Plane Microstrip Patch Antenna Using Plastic Deformation Magnetic Actuation," IEEE Microwave and Wireless Components Letters, vol. 13, no. 3, pp. 120–122, March 2003. DOI: 10.1109/LMWC.2003.810123. 25, 26

[84] B. Poussot, J. M. Laheurte, L. Cirio, O. Picon, D. Delcroix, and L. Dussopt, "Diversity measurements of a reconfigurable antenna with switched polarizations and patterns," IEEE Transactions on Antennas and Propagation, vol. 56, Issue 1, pp. 31–38, Jan. 2008. DOI: 10.1109/TAP.2007.913032. 26, 27, 28

[85] J. Costantine, S. al-Saffar, C. G. Christodoulou, K. Y. Kabalan, A. El-Hajj, "The Analysis of a Reconfigurable Antenna With a Rotating Feed Using Graph Models," IEEE Antennas and Wireless Propagation Letters, Accepted for Future Publication, vol. PP, pp. 1–1, 2009. DOI: 10.1109/LAWP.2009.2029137. 29, 30

[86] J. A. Bossard, D. H. Werner, T. S. Mayer, and R. P. Drupp, "A Novel Design Methodology for Reconfigurable Frequency Selective Surfaces Using Genetic Algorithms," IEEE Transactions on Antennas and Propagation, vol. 53, Issue 4, pp. 1390–1400, April 2005. DOI: 10.1109/TAP.2005.844439. 33, 49

[87] D. Ressiguier, J. Costantine, Y. Tawk, and C. G. Christodoulou, "A Reconfigurable Multi-Band Antenna Based on Open Ended Microstrip Lines," Third European Conference on Antennas and Propagation 2009, pp. 792–795, March 2009. 34, 35, 38, 39, 57

[88] D. E. Anagnostou, Z. Guizhen, M. T. Chrysomallis, J. C. Lyke, G. E. Ponchak, J. Papapolymerou, and C. G. Christodoulou, "Design, fabrication and measurement of an RF-MEMS-based self–similar reconfigurable antenna," IEEE Transactions on Antennas and Propagation, vol. 54, Issue 2, Part 1, pp. 422–432, Feb 2006. DOI: 10.1109/TAP.2005.863399. 35

[89] A. Patnaik, D. E. Anagnostou, C. G. Christodoulou, and J. C. Lyke, "A frequency reconfigurable antenna design using neural networks," IEEE Antennas and Propagation society international symposium, vol. 2A, pp. 409–412, July 2005. DOI: 10.1109/APS.2005.1551830. 49

[90] N. Kingsley, D. E. Anagnostou, M. Tentzeris, and J. Papapolymerou, "RF MEMS sequentially reconfigurable sierpinski antenna on a flexible organic substrate with novel DC-biasing technique," Journal of microelectromechanical systems, vol. 16, Issue 5, pp. 1185–1192, Oct. 2007. DOI: 10.1109/JMEMS.2007.902462.

[91] F. Ghanem, P. S. Hall, and J. R. Kelly, "Two port frequency reconfigurable for cognitive radios," IEEE Electronic Letters, vol. 45, Issue 11, pp. 534–536, May 2009. DOI: 10.1049/el.2009.0935.

[92] A. C. K. Mak, C. R. Rowell, R. D. Murch, and C. L. Mak, "Reconfigurable multiband antenna designs for wireless communication devices," IEEE Transactions on Antennas and Propagation, vol. 55, Issue 7, pp. 1919–1928, July 2007. DOI: 10.1109/TAP.2007.895634.

[93] A. Cidronali, L. Lucci, G. Pelosi, P. Sarnori, and S. Selleri, "A reconfigurable printed dipole for quad-band wireless applications," IEEE Antennas and Propagation society international symposium, pp. 217–220, July 2006. DOI: 10.1109/APS.2006.1710493.

[94] M. A. Ali and P. Waheed, "A reconfigurable yagi-array for wireless applications," IEEE Antennas and Propagation society international symposium, vol. 1, pp. 446–468, June 2002. DOI: 10.1109/APS.2002.1016385.

[95] L. M. Feldner, C. T. Rodenbeck, C. G. Christodoulou, and N. Kinzie, "Electrically small frequency-agile PIFA-as-a-package for portable wireless devices," IEEE Transactions on Antennas and Propagation, vol. 55, Issue 11, pp. 3310–3319, Nov. 2007. DOI: 10.1109/TAP.2007.908815.

[96] W. H. Weedon, W. J. Payne, and G. M. Rebeiz, "MEMS-switched reconfigurable antenna," IEEE Antennas and Propagation society symposium, vol. 3, pp. 654–657, July 2001. DOI: 10.1109/APS.2001.960181.

[97] S. Xiao, B. Z. Wang, X. S. Yang, and G. Wang, "Reconfigurable microstrip antenna design based on genetic algorithm," IEEE Antennas and Propagation society symposium, vol. 1, pp. 407–410, June 2003. DOI: 10.1109/APS.2003.1217483.

[98] V. Zachou, C. G. Christodoulou, M. T. Chrisomallis, D. Anagnostou, and S. E. Barbin, "Planar Monopole Antenna With Attached Sleeves," Antennas and Wireless Propagation Letters, vol. 5, no. 1, pp. 286–289, Dec. 2006. DOI: 10.1109/LAWP.2006.876970. 42, 43, 44

[99] J. M. Zendejas, J. P. Gianvittorio, Y. Rahmat-Samii, and J. W. Judy, "Magnetic MEMS Reconfigurable Frequency Selective Surfaces," Journal of Microelectromechanical Systems, vol. 15, no. 3, pp. 613–623, June 2006. DOI: 10.1109/JMEMS.2005.863704.

[100] J. Costantine and C. G. Christodoulou, "A new reconfigurable antenna based on a rotated feed," IEEE international symposium on antennas and propagation, pp. 1–4, July 2008. DOI: 10.1109/APS.2008.4619618.

[101] J. Costantine, Y. Tawk, C. G. Christodoulou, and S. E. Barbin, "A Star Shaped Reconfigurable Patch Antenna," IEEE MTT-S International Microwave Workshop Series on Signal Integrity and High Speed Interconnects, pp. 97–100, 2009. DOI: 10.1109/IMWS.2009.4814917. 59, 60, 76, 77, 78

[102] M. T. Oswald, S. C. Hagnes, B. D. Van Veen, and Z. Popovic, "Reconfigurable single-feed antennas for diversity communications," IEEE Antennas and Propagation society international symposium, vol. 1, pp. 469–472, June 2002. DOI: 10.1109/APS.2002.1016386.

[103] H. Aissat, L. Cirio, M. Grzeskwiak, J. M. Laheurte, and O. Picon, "Reconfigurable circularly polarized antenna for short-range communication systems," IEEE Transactions on Microwave Theory and Techniques, vol. 54, Issue 6, Part 2, pp. 2856–2863, June 2006. DOI: 10.1109/TMTT.2006.875454.

[104] F. Yang and Y. Rahmat Samii, "Patch antenna with switchable slots (PASS): reconfigurable design for wireless communications," IEEE Antennas and Propagation Society International Symposium, vol. 1, pp. 462–465, 2002. DOI: 10.1109/APS.2002.1016384.

[105] Y. Tawk, C.G. Christodoulos, and J. Costantine, "Radiation and Frequency Reconfiguration Using Tilted Printed Monopoles," IEEE International Symposium on Antennas and Propagation, 2013.

[106] C. G. Christodoulou, J. H. Kim, J. Costantine, and S. E. Barbin, "Reconfigurable RF and Antenna Systems," SBMO/IEEE MTT-S International Microwave and Optoelectronics Conference, pp. 17–20, Nov. 2007. DOI: 10.1109/IMOC.2007.4404203. 35

[107] Z. Min, L. Xiao-Wu, and W. Guang-Hui, "Preliminary Research of the Reconfigurable Antenna Based on Genetic Algorithms," 2004 third International Conference on Computational ELectromagnetics and its Applications, pp. 137–140, Nov. 2004. DOI: 10.1109/ICCEA.2004.1459309. 49

[108] D. E. Skinner, J. D. Connor, S. Y. Foo, M. H. Weatherspoon, and N. Powell, "Optimization of Multi-Band Reconfigurable Microstrip Line-Fed Rectangular Patch Antenna Using Self-Organizing Maps," IEEE 10th Annual Wireless and Microwave Technology Conference (WAMICON), pp. 1–4, Apr. 2009. DOI: 10.1109/WAMICON.2009.5207263. 49

[109] A. Akdagli, K. Guney, and B. Babayigit, "Clonal Selection Algorithm for Design of Reconfigurable Antenna Array with Discrete Phase Shifters," Journal of Electromagnetic Waves and Applications, vol. 21, no. 2, pp. 215–227, 2007. DOI: 10.1163/156939307779378808. 49

[110] C. M. Coleman, J. E. Rothwell, and J. E. Ross, "Investigation of Simulated Annealing, Ant-Colony and Genetic Algorithms for Self-Structuring Antennas," IEEE Transactions on Antennas and Propagation, vol. 52, Issue 4, pp. 1007–1014, Apr. 2004. DOI: 10.1109/TAP.2004.825658. 50

[111] J. D. Connor, "Antenna Array Synthesis Using the Cross Entropy Method," Ph.D. Dissertation, Florida State University, Tallahassee, Fl., USA, June 2008. 50

[112] D. Langoni, M. H. Weatherspoon, E. Ogunti, and S. Y. Foo, "An Overview of Reconfigurable Antennas: Design, Simulation, and Optimization," IEEE 10th Annual Wireless and Microwave Technology Conference (WAMICON), pp. 1–5, Apr. 2009. DOI: 10.1109/WAMICON.2009.5207264. 50

[113] J. Costantine, S. al-Saffar, C. G. Christodoulou, and C. T. Abdallah, "Reducing Redundancies in Reconfigurable Antenna Structures Using Graph Models," IEEE Transactions on Antennas and Propagation, vol. 59, Issue 3, pp. 793–801, 2011. DOI: 10.1109/TAP.2010.2103005. 50, 51, 61, 62, 65, 69, 70, 76

[114] V. Balakrishnan, "Schaum's Outline of Graph Theory: Including Hundreds of Problems," McGraw- Hill Companies Inc., 1997. 52

[115] A. Grau, L. Ming-Jer, J. Romeu, H. Jafarkhani, L. Jofre, and F. De Flaviis, "A Multifunctional MEMS-Reconfigurable Pixel Antenna For Narrowband MIMO Communications," IEEE Antennas and Propagation society international symposium , pp. 489–492, June 2007. DOI: 10.1109/APS.2007.4395537. 61, 63, 64, 65, 71

[116] R. G. Vaughan, "Two-port higher mode circular microstrip antennas," IEEE Transactions on Antennas and Propagation, vol. 36, Issue 3, pp. 309–321, Mar. 1988. DOI: 10.1109/8.192112. 63

[117] J. Costantine, Y. Tawk, C. G. Christodoulou, J. C. Lyke, F. De Flaviis, A. Grau Besoli, and S. E. Barbin, "Analyzing the Complexity and Reliability of Switch-Frequency-Reconfigurable Antennas Using Graph Models," IEEE Transactions on Antennas and Propagation, vol. 60, Issue 2, Part 2, pp. 811–820, 2012. DOI: 10.1109/TAP.2011.2173104. 75, 76, 77, 78, 81, 85, 86, 87, 96

[118] H. Chang, J. Quian, B. A. Cetiner, F. De Flaviis, M. Bachman, and G. P. Li, "RF MEMS Switches Fabricated on Microwave-Laminate Printed Circuit Boards," IEEE Electron Device Letters, vol. 24, Issue 4, pp. 227–229, April 2003. DOI: 10.1109/LED.2003.812150. 75

[119] C. W. Jung and F. De Flaviis, "RF-MEMS Capacitive Series of CPW&MSL Configurations for Reconfigurable Antenna Application," IEEE Antennas and Propagation Society International Symposium, vol. 2A, pp. 425–428, July 2005. DOI: 10.1109/APS.2005.1551834.

[120] G. Wang, T. Polley, A. Hunt, and J. Papapolymerou, "A High Performance Tunable RF MEMS Switch Using Barium Strontium Titanate (BST) Dielectrics for Reconfigurable Antennas and Phased Arrays," IEEE Antennas and Wireless Propagation Letters, vol. 4, pp. 217–220, 2005. DOI: 10.1109/LAWP.2005.851065. 75

132 BIBLIOGRAPHY

[121] D. E. Anagnostou, G. Zheng, M. Chryssomallis, J. Lyke, G. Ponchak, J. Papapolymerou, and C. G. Christodoulou, "Design, Fabrication and Measurements of an RF-MEMS-Based Self-Similar Reconfigurable Antenna," IEEE Transactions on Antennas & Propagation, Special Issue on Multifunction Antennas and Antenna Systems, vol. 54, Issue 2, Part 1, pp. 422–432, Feb. 2006. DOI: 10.1109/TAP.2005.863399. 75

[122] A. Carton, C. G. Christodoulou, C. Dyck, and C. Nordquist, "Investigating the impact of Carbon Contamination on RF MEMS Reliability," IEEE Antennas and Propagation International Symposium, pp. 193–196, July 2006. DOI: 10.1109/APS.2006.1710487. 75, 91

[123] E. F. Moore and C. E. Shannon, "Reliable Circuits Using Less Reliable Relays," (Part I and Part II), J. Franklin Inst., vol. 262, no. 3 (Sept. 1956), pp. 191–208 and no. 4 (Oct. 1956), pp. 281–297. DOI: 10.1016/0016-0032(56)90559-2. 75, 79, 84

[124] N. Pippenger and G. Lin, "Fault-Tolerant Circuit-Switching Networks," SIAM Journal on Discrete Mathematics, vol. 7, Issue 1, pp. 108–118, June 1994. DOI: 10.1137/S0895480192229790. 75

[125] J. Costantine and C. G. Christodoulou, "Analyzing Reconfigurable Antenna Structure Redundancy Using Graph Models," IEEE Antennas and Propagation Society International Symposium 2009, pp. 1–4, June 2009. DOI: 10.1109/APS.2009.5171562. 76

[126] J. Costantine, C. G. Christodoulou, C. T. Abdallah, and S.E. Barbin, "Optimization and Complexity Reduction of Switch-Reconfigurable Antennas Using Graph Models," IEEE Antennas and Wireless Propagation Letters, vol. 8, pp. 1072–1075, 2009. DOI: 10.1109/LAWP.2009.2032674.

[127] J. Costantine, Y. Tawk, C. G. Christodoulou, and C. T. Abdallah, "Reducing Complexity and Improving the Reliability of Frequency Reconfigurable Antennas," 2010 Proceedings of the Fourth European Conference on Antennas and Propagation, pp. 1–4, 2010. 76

[128] A. M. Sayeed and V. Raghavan, "On the impact of reconfigurable antenna arrays in cognitive radio," IEEE International Conference on Acoustics, Speech and Signal processing, vol. 4, pp. 1353–1356, 2007. DOI: 10.1109/ICASSP.2007.367329. 91

[129] P. Mookiah, D. Piazza, and K. R. Dandekar, "Reconfigurable Spiral Antenna Array for Pattern Diversity in Wideband MIMO Communication systems," IEEE International Symposium on Antennas and Propagation, pp. 1–4, 2008. DOI: 10.1109/APS.2008.4619210. 91

[130] M. T. Ali, T. A. Rahman, M. R. Kamarudin, R. Sauleau, M. N. M. Tan, and M. F. Jamlos, "A Reconfigurable Planar Antenna Array (RPAA) With Back Lobe Reduction," International Workshop and Antenna Technology (IWAT), pp. 1–4, 2010. DOI: 10.1109/IWAT.2010.5464784. 91, 92

[131] R. J. Chaffin, "Thermally Induced Switching and Failure in p-i-n RF Control Diodes," IEEE Transactions on Microwave Theory and Techniques, vol. 30, no. 11, pp. 1944–1947, Nov. 1982. DOI: 10.1109/TMTT.1982.1131348. 91

[132] J. Costantine, Y. Tawk, and C. G. Christodoulou, "Complexity Versus Reliability in Arrays of Reconfigurable Antennas," IEEE Transactions on Antennas and Propagation, vol. 60, Issue 11, pp. 5436–5441, 2012. DOI: 10.1109/TAP.2012.2207665. 92, 93, 94, 95, 96, 97, 100, 101, 102

[133] Renesas Electronic Corporation, "Diode: Reliability of Renesas Diodes," April 1, 2010. http://documentation.renesas.com/doc/products/diode/rej27g0006_diode.pdf 98

[134] J. G. Teti and F. P. Dareff, "MEMS 2-bit Phase Shifter Failure Mode and Reliability Considerations For Large X Band Arrays," IEEE Transactions on Microwave Theory and Techniques, vol. 52, Issue 2, pp. 693–701. DOI: 10.1109/TMTT.2003.822017. 99, 101

[135] Philips Semiconductors, "Data Sheet BAP64-03 Silicon PIN Diode," February 11, 2004. http://datasheet.octopart.com/BAP64--03-T/R-Philips-datasheet-9082.pdf 102

[136] Y. Cai and Z. Du, "A Novel Pattern Reconfigurable Antenna Array for Diversity Systems," IEEE Antennas and Wireless Propagation Letters, vol. 8, pp. 1227–1230. DOI: 10.1109/LAWP.2009.2035720. 102

[137] J. De Luis and F. De Flaviis, "Frequency Agile Switched Beam Antenna Array," IEEE Transactions on Antennas and Propagation, vol. 58, Issue 10, pp. 3196–3204, 2010. DOI: 10.1109/TAP.2010.2055813. 102

[138] Macom Technology Solutions, "Package PIN Diodes," 2013. http://www.macomtech.com/datasheets/packagedpindiodes.pdf 102

[139] Macom Technology Solutions, "Application Notes S2080: Moisture Effects on the Soldering of Plastic Encapsulated Devices," 2013. http://www.macomtech.com/Application%20Notes/pdf/S2080.pdf 102

[140] J. A. Rodriguez, M. Fernandez-Delgado, J. Bregains, R. Iglesias, S. Barro, and F. Ares, "A Comparison Among Several Techniques for Finding Defective Elements in Antenna Arrays," The Second European Conference on Antennas and Propagation, EuCAP 2007, pp. 1–8, 11–16 Nov. 2007. 105

[141] M. J. Rivera, "Failure Detection and Correction in Switch Reconfigurable Antenna Arrays," Master's Thesis, 2011. DOI: 10.1109/APS.2011.5996441. 106, 107, 109, 111, 114, 118

[142] M. J. Rivera, J. Costantine, Y. Tawk, and C. G. Christodoulou, "Failure Detection and Correction in Switch Reconfigurable Antenna Arrays," IEEE International Symposium on Antennas and Propagation, 2011. DOI: 10.1109/APS.2011.5996441. 106, 107, 109, 116, 117

[143] M. J. Rivera, J. Costantine, Y. Tawk, and C. G. Christodoulou, "Detection of Failures in Switch Reconfigurable Antenna Arrays Using Embedded Sensing Lines," IEEE International Symposium on Antennas and Propagation, 2012. DOI: 10.1109/APS.2012.6347989. 108, 110, 111, 112, 113

[144] J. Costantine, M. J. Rivera, Y. Tawk, and C. G. Christodoulou, "Overcoming Failures in Reconfigurable Antenna Arrays Using Equivalent Frequency Dependent Graphs," Proceedings of the 5th European Conference on Antennas and Propagation, pp. 2152–2155, 2011. 115, 116

Authors' Biographies

JOSEPH COSTANTINE

Joseph Costantine received his Ph.D. degree from the University of New Mexico in 2009 where he also completed a post-doc fellowship in July 2010. Dr. Costantine has a Masters in Computer and Communications Engineering from the American University of Beirut in 2006. His bachelor's degree is in Electrical, Electronics, Computer, and Communications Engineering from the second branch of the Faculty of Engineering in the Lebanese University in 2004. He is the recipient of many awards including the summer faculty fellowship from the space vehicles directorate in Albuquerque, NM, for three consecutive years. He has published many research papers and has several patents. His major research interests reside in reconfigurable antennas for wireless communication systems, cognitive radio, antennas for biomedical applications, and deployable antennas for small satellites.

YOUSSEF TAWK

Youssef Tawk received the Ph.D. degree from the University of New Mexico in 2011 where he also completed a post-doc fellowship in August 2012. Dr. Tawk has a Master's degree in Electrical and Computer Engineering from the American University of Beirut in 2008. His bachelor's degree in Computer and Communication Engineering was received with highest distinction in 2006 from Notre Dame University, Louaize, Lebanon. He received many awards and honors throughout his studies. He has published several journal and conference papers and has several issued patents. His research interests include reconfigurable antenna systems, cognitive radio, radio-frequency (RF) electronic design, photonics, and millimeter-wave technology.

CHRISTOS G. CHRISTODOULOU

Christos G. Christodoulou received his Ph.D. degree in Electrical Engineering from North Carolina State University in 1985. He served as a faculty member in the University of Central Florida, Orlando, from 1985 to 1998. In 1999, he joined the faculty of the Electrical and Computer Engineering Department of the University of New Mexico, where he served as the Chair of the Department from 1999 to 2005. He is a Fellow member of IEEE and a member of Commission B of URSI. Currently he is the director for COSMIAC (Configurable Space Microsystems Innovations & Applications Center at UNM). He is the recipient of the 2010 IEEE John Krauss Antenna Award, the Lawton-Ellis Award, and the Gardner Zemke Professorship at the University of New Mexico. He has published about 400 papers in journals and conferences, has 13 book chapters, and has coauthored five books. His research interests are in the areas of modeling of electromagnetic systems, adaptive array antennas, high power microwave antennas, reconfigurable antenna systems, cognitive radio, and smart RF/photonics.

Printed in the United States
by Baker & Taylor Publisher Services